The Insectile and the Deconstruction of the Non/Human

The Insectile and the Deconstruction of the Non/Human defines, conceptualizes, and evaluates the insectile—pertaining to an entomological fascination—in relation to subject formation. The book is driven by a central dynamic between form and formlessness, further staging an investigation of the phenomenon of fascination using Lacanian psychoanalysis, suggesting that the psychodrama of subject formation plays itself out entomologically. The book's engagement with the insectile—its enactments, cultural dreamwork, fantasy transformations—'in-forming' the so-called human subject undertakes a broader deconstruction of said subject and demonstrates the foundational but occluded role of the insectile in subject formation. It tracks the insectile across the archives of psychoanalysis, seventeenth century still life painting, novels from the nineteenth century to the present day, and post-1970s film. *The Insectile and the Deconstruction of the Non/Human* will be of interest for scholars, graduate students, and upper-level undergraduates in film studies, visual culture, popular culture, cultural and literary studies, comparative literature, and critical theory, offering the insectile as new category for theoretical thought.

Fabienne Collignon is Senior Lecturer in Contemporary Literature at the University of Sheffield, UK. Her research interests are critical theory, in particular theories of technology, subject formation, the 'in-human'. She has published articles in *Textual Practice*, *C-Theory*, *Journal of American Studies*, *Orbit*, *Configurations, New Formations*. Her first monograph, *Rocket States: Atomic Weaponry and the Cultural Imagination*, was published by Bloomsbury in 2014 and maps the technological unconscious of the Cold War. From September 2018 to February 2020, she was an Alexander von Humboldt Research Fellow at Universität zu Köln, during which time she researched scenes of the insectile.

Perspectives on the Non-Human in Literature and Culture

Series Editor: Karen Raber, *University of Mississippi, USA*

Literary and cultural criticism has ventured into a brave new world in recent decades: posthumanism, ecocriticism, critical animal studies, the new materialisms, the new vitalism, and other related approaches have transformed the critical environment, reinvigorating our encounters with familiar texts, and inviting us to take note of new or neglected ones. A vast array of non-human creatures, things, and forces are now emerging as important agents in their own right. Inspired by human concern for an ailing planet, ecocriticism has grappled with the question of how important works of art can be to the preservation of something we have traditionally called "nature." Yet literature's capacity to take us on unexpected journeys through the networks of affiliation and affinity we share with the earth on which we dwell—and without which we die—and to confront us with the drama of our common struggle to survive and thrive has not diminished in the face of what Lyn White Jr. called "our ecological crisis." From animals to androids, non-human creatures and objects populate critical analyses in increasingly complex ways, complicating our conception of the cosmos by dethroning the individual subject and dismantling the comfortable categories through which we have interpreted our existence. Until now, however, the elements that compose this wave of scholarship on non-human entities have had limited places to gather to be nurtured as a collective project. "Perspectives on the Non-Human in Literature and Culture" provides that local habitation. In this series, readers will find creatures of all descriptions, as well as every other form of biological life; they will also meet the non-biological, the microscopic, the ethereal, the intangible. It is our goal for the series to provide an encounter zone where all forms of human engagement with the non-human in all periods and national literatures can be explored, and where the discoveries that result can speak to one another, as well as to scholars and students.

Entangled Fictions
Nonhuman Animals in an Indian World
Suvadip Sinha

Futures of the Human Subject
Technical Mediation, Foucault and Science Fiction
Sławomir Kozioł

Representing Post(Human) Enhancement Technologies in Twenty-First Century US Fiction
Carmen Laguarta-Bueno

For more information about this series, please visit: www.routledge.com/Perspectives-on-the-Non-Human-in-Literature-and-Culture/book-series/PNHLC

The Insectile and the Deconstruction of the Non/Human

Fabienne Collignon

Routledge
Taylor & Francis Group
NEW YORK AND LONDON

Cover image: © Jorg Greuel / Getty Images

First published 2023
by Routledge
605 Third Avenue, New York, NY 10158, USA

and by Routledge
4 Park Square, Milton Park, Abingdon, Oxon, OX14 4RN, UK

Routledge is an imprint of the Taylor & Francis Group, an informa business.

Library of Congress Cataloging-in-Publication Data
Names: Collignon, Fabienne, author.
Title: The insectile and the deconstruction of the non/human / Fabienne Collignon.
Description: New York, NY : Routledge, 2023. | Series: Perspectives on the non-human in literature and culture | Includes bibliographical references and index. |
Identifiers: LCCN 2022036649 (print) | LCCN 2022036650 (ebook) | ISBN 9781032345505 (hardback) | ISBN 9781032345512 (paperback) | ISBN 9781003322757 (ebook)
Subjects: LCSH: Insects in literature. | Insects--Symbolic aspects. | Other (Philosophy) in literature. | Insects in art. | Other (Philosophy) in art.
Classification: LCC PN56.I63 C65 2023 (print) | LCC PN56.I63 (ebook) | DDC 809/.9336257--dc23/eng/20221011
LC record available at https://lccn.loc.gov/2022036649
LC ebook record available at https://lccn.loc.gov/2022036650

ISBN: 978-1-032-34550-5 (hbk)
ISBN: 978-1-032-34551-2 (pbk)
ISBN: 978-1-003-32275-7 (ebk)

DOI: 10.4324/9781003322757

Typeset in Sabon
by SPi Technologies India Pvt Ltd (Straive)

Contents

Acknowledgements

I would, above all, like to thank the Alexander von Humboldt Foundation for their generous grant in support of this project, which enabled me to conduct my research in Germany. The grant made me rediscover the pleasures of research and writing and to combine these pleasures with other passions, too, such as Borussia Dortmund football games ('Nur der BVB!') and living in Köln and Berlin, in spaces of support and safety, given over to the enjoyment of the No (thank you, Hannah Hoffmann in K and Heather Purcell in B). I would also like to thank the people I worked with at Universität zu Köln, first and foremost Hanjo Berressem, as well as Roman Bartosch, Sandro Eich, Bärbel Eltschig, Nina Engelhardt, Julia Hoydis and Jonas Nelder; salute the people I met through the Humboldt network, notably Laura Jarmatz and Kathrin Yacavone; and thank Andreas Kahrs, not only for his expertise and reading suggestions on the historical and contemporary German far right and his hair-cutting skills but also for the company (invoking Johannes Ehmann on this note, too) during the pandemic, watching the otherwise unbearable *Geisterspiele*. I want to further thank colleagues at the University of Sheffield and elsewhere who read drafts of this project—Matthew Cheeseman, Alex Dunst, Seb Franklin, Michael Kindellan, Rachel Murray, Adam Piette, Sue Vice and Maisha Wester—as well as my PhD students, especially Greg Keenan, Sophie Maxwell, Pete Sands and Rich Woodall, for their inspiration. Finally, I want to extend my profound gratitude to everyone I love in the places I have lived, visited, keep returning to and/or continue to live.

1 Introduction

Insectile Subjectile

Scenes of Fascination: I

In Thomas Hardy's *The Return of the Native* (1878), the illicit relationship between Wildeve and Eustacia is figured in terms of the trope of the moth propelled into the flame of a candle. This trope, on the one hand, functions as communication signal between Wildeve and Eustacia: to alert her, inside, to his presence by her window outside, Wildeve captures a moth to release it close to a candle's light source, whereupon the insect burns itself up. On the other, it behaves as metaphor for their fatal attraction to each other, the manoeuvre further pointing to their insensitivity toward (non-human) others. In its plurality of meanings, this means of communication manages to transmit the shock or short circuit of their love affair, the inability to navigate the insistent force of desire:

> The heath tonight appeared to be totally deserted: and Wildeve, after looking over Eustacia's garden gate for some time, with a cigar in his mouth, was tempted by the fascination that emotional smuggling had for his nature to advance towards the window, which was not quite closed, the blind being only partly drawn down. He could see into the room, and Eustacia was sitting there alone. Wildeve contemplated her for a minute, and then retreating into the heath beat the ferns lightly, whereupon moths flew out alarmed. Securing one, he returned to the window, and holding the moth to the chink, opened his hand. The moth made towards the candle upon Eustacia's table, hovered around it two or three times, and flew into the flame.[1]

The mechanism at work here is fascination, a phenomenon that structures much of Hardy's novel (the third section is titled 'Fascination'), concerning not only Wildeve and Eustacia's relationship but also the encounter between Clym Yeobright and Eustacia or, more generally, the subject's relationship to idealised objects. Clym's face, for example, forms a singular element of attention in the narrative, to the point that it appears as 'typical countenance of the future' according to the omniscient

DOI: 10.4324/9781003322757-1

narrator, remarking that an observer's eye is 'arrested' by Yeobright's face not as picture or image but as page, by the language it records.[2] In the lineaments of Clym's face, the narrator discerns or projects a trace of the future, in other words, the conditions of possibility of a 'type' of subject. Clym's features are 'attractive in the light of symbols, as sounds intrinsically common become attractive in language, and as shapes intrinsically simple become interesting in writing.'[3] The narrator observes a process of forming, of formless 'things' (sounds; shapes) shifting into forms (symbols; language; writing) taking place in Clym's face which, though 'page,' is also an (arresting; fascinating) image. The narrator acknowledges as much by referring to Pheidias' production of heads, the 'language' of their idealised renditions in high classical art.

The scene of fasciation in the novel, as such, spreads beyond its most compelling and tragic evocation, illustrating Wildeve's orientation towards Eustacia and vice versa. The subsection titled 'Fascination' begins with the 'system' of Clym's face and figure, moving through a series of encounters and settings in which the phenomenon takes hold or effect. The metaphor of the moth and its attraction to light—the event of fascination *par excellence*—frequently exists latently, in a shared glance between Clym and Eustacia coded in terms of a 'moonlit scene,' occurring in broad daylight, 'common to both'.[4] Moths are primarily nocturnal creatures travelling by the light of the moon, kept in a precise location in relation to their bodies. The mutual glance (Clym→ ←Eustacia) takes place in a field of signification that draws on this behaviour, intimated in one body's transverse orientation toward the other. In this instance, both bodies, held in the light and gaze of the other, are compelled toward one another, each of them at once moth and moon. The shared glance, the capture in the other's gaze, constitutes the scene of fascination, in itself demonstrating how the process works to entrance, to lose one's head and sense of time and place: in an instant, the world disappears.

The nodes of the current project become apparent in *The Return of the Native* in its constellation of insects (specifically moths), fascination and subjectivity. This constellation has multiple expressions and patterns in and outside the novel, like the conditions of light or of the gaze arresting the subject, caught in a behavioural convergence with moths. The aim of this book is to demonstrate the foundational but occluded role of the insectile in subject formation; to do so, it brings together and evaluates a library of writings on, or renditions of, the insectile in relation to the production of subjectivity. The insectile is that which prompts a rethinking of the so-called human subject's enunciation, figured and unfigured through the phenomenon of fascination. Before, however, mapping this realm of fascination—the main thinkers on this matter are Jacques Lacan and Maurice Blanchot—I want to lay out the other central relation that organises what is to come, all the while keeping the state of fascination in mind, but some way off, at the edges of awareness. One reason for this decentring is to be found in the phenomenon of fascination itself, which leads the I elsewhere, into a dimension where 'all subjective subsistence seems to get lost, to be absorbed, and to leave the world behind'.[5] Another motive for circumventing it here has to do with its consequences for the ego, which is at once formed and deformed by being fascinated: fascination brings the subject into existence at the same time that it is moved outside or beyond itself. The evocation of both form and formlessness pertains to the scene of fascination and, as such, to the constitution of the so-called human subject; this conceptual frame (form; formlessness; fascination) is what drives this investigation. Form and formlessness are interlinked, as are fascination and insects, creatures that fascinate, as if they were manifestations of the gaze itself. In *Flights* (2018), Olga Tokarczuk writes that an enormous black beetle, whose 'flawless carapace' reflects the sky, functions as an 'odd eye on the ground, not belonging to any body, detached and disinterested'.[6]

I am using Hardy's novel to initiate an investigation into the insectile and *Un/forms* of fascination to rethink the notion of the so-called human subject in-formed, internally formed, by the insectile trace of the other. In a first instance, my methodology brings Jacques Derrida into relation with Lacan: the concept of the insectile arises out of their work. The insectile occurs in Lacanian thought, notably concerning the Ideal-I, the projected image of the subject, and *objet a*, the dimension of the other's desire which prompts the subject's constitution, rendering itself into an imago which assumes the armour of subjectivity. The insectile, further, maps onto the Möbius strip, with which Lacan illustrates the topography of the subject and which in *Seminar X: Anxiety* (1962–1963) is crawling with ants (Figure 1.1): right from the start of its invocation, the subject is a structure or surface implicated in the insectile. The Möbius strip organises the movements of the insectile as it twists into and out of form: by form, I refer to the Ideal-I, to fictions of armoured subjectivity that emerge according to an entomological imagination. In *Vampyroteuthis Infernalis* (1987), the German media theorist Vilém Flusser stages a game of reflection between 'us,' so-called humans, and a species of octopods difficult to classify: *vampyroteuthis infernalis*, discovered in the South China Sea, which despite living 'far apart' is 'not entirely alien to us'.[7] Flusser's proposal, a fable and metaphorical enterprise, functions to release 'us' into the domain of the absolute other and constitutes a process of emancipation, an emergence from one order—vertebrate, linear ('we think linearly')—into a molluscan, 'eccentric' condition of possibility.[8] He thereby seeks to destabilise the genre of the 'human' as abyssal, octopodal, and while Flusser's object of analysis (*vampyroteuthis*) is not my own, his engagement with the 'form' of the 'human,' cast into disarray, helps organise my own investigation on said form, intertwined, as it is, with the insectile.

Saying that, there are significant differences between Flusser's fable and the current project, not only with respect to its studied object. Just because a transformation is taking place as a result of a game of reflection between the I, so-called human subject, and an Other does not mean that the discourse of the 'human'—an 'abstract' citizen, to refer to Lauren Berlant—as the good form is disrupted.[9] If the insectile has to do with the enunciation of the so-called human, that 'in-forming' enunciation does not necessarily yield a radical politics but can also function as a system of meaning stabilising around an exalted and resplendent phallic subject. I insist, then, on the turns of the insectile, topologically a Möbius strip, as both potentially form-resistant and an absolute mobilisation of the self, a state of things that I trace throughout this study and that can be glimpsed, by way of an example, in Colson Whitehead's 2009 novel *Sag Harbour*. The novel follows Benji Cooper, a black teenager spending the summer of 1985 in a village in the Hamptons, working in an ice cream shop called Jonny Waffle. Benji—who generally gets the impression that insects refer to 'elusive [profundities],' not least concerning the correlations between them and racialised abjection or dispossession—reflects on fireflies:

> A firefly blinked into existence, drew half a word in the air. Then gone. A black bug secret in the night. Such a strange little guy. It materialised, visible to human eyes for brief moments, and then it disappeared. But it got its name from its fake time, people time, when in fact most of its business went on when people couldn't see it. Its true life was invisible to us but we called it firefly after its fractions. Knowable and fixed for a few seconds, sharing a short segment of its message before it continued on its real mission, unknowable in its true self and course, outside of reach. It was a bad name because it was incomplete—both parts were true, the bright and the dark, the one we could see and the other one we couldn't. It was both.[10]

Benji thinks of the insect in terms of fractions, episodes of visibility and invisibility, 'fake time' and 'true life,' knowable and unknowable segments, an entity of two parts drawing half or incomplete words 'in the air' by virtue of its existence. I extrapolate Benji's reflection to include form and formlessness, there between the lines, in those aspects that disappear outside 'people time' and so-called human thought, though formlessness does not invariably or automatically shatter language, established norms nor the socio-political and economic order. In 'The Context of Forms,' Seb Franklin shows how formlessness is produced so as to sustain the logic of racial capitalism, establishing form as value and the formless as racialised abjection to be captured, exploited and discarded.[11] This conceptualisation of form and formlessness, as it pertains to subject formation and representation as well as to social and political life, structures what is to come, thereby also staging a critique of Gilles Deleuze and Félix Guattari's notion of becoming developed in *A Thousand Plateaus* (1980).

Any investigation of (ostensible) becoming-other has to refer to Deleuze and Guattari's work. The insectile, as concept and process, cites becoming-insect but remains vigilant as to its operations. The 'scandal' of this vigilance involves Lacan, brought into play to think through the movements between formless assemblages and the Ideal-I. This *point de rencontre*, between Lacan and Deleuze and Guattari, is set up through Derrida, the idea of the missed encounter and his practice of reading, writing and interpretation as itself insectile, parasitic.

Lacan's thinking is central to this book: his work unlocks the overlooked centrality of the insectile in theories of the subject and allows us to conceptualise the insectile in its turns between form and formlessness. Lacanian psychoanalysis is something few posthumanities scholars are willing to contemplate, thereby failing to notice the precarity of the so-called human subject at the moment it is called into being; missing the connections between entomology, fascination and subjectivity; replicating and perpetuating models of thought forbidding the encounter between Lacan and Deleuze and Guattari. By using Derrida's work, we can, however, imagine other encounters between these thinkers and thereby show how Lacanian subject formation locates the insectile—as in-forming posthumanism—right at the 'heart' of the subject.

The book's argument moves through the insectile implications of the so-called human and the changing, rather than ontological, states of form and formlessness; its aims are to contribute to debates about an 'in-forming' posthumanism and to follow the (political) dimensions of the posthuman in its insectile 'forms'. The theoretical material accumulates in this introduction, occasionally referring to other cultural texts—the insectile, after all, hosts multitudes—all the while structuring the remaining chapters, revolving around the event of fascination and constructions of form and formlessness through the 'unholy' alliance of Lacan with Deleuze and Guattari. Rather than unholy, though, this alliance is under-theorized. *Anti-Oedipus*, written in 1968 and published in 1972, for example, is as much a 'work *of* psychoanalysis,' according to Dagmar Herzog, than a critique of a particular psychoanalytic practice taking place in a Cold War context that, above all in the US, severed the relationship between the subject and social structures.[12] In this vein, the argument below performs that which is long overdue: to put Lacan to work with Deleuze and Guattari. Iterations of the insectile—often arriving unexpectedly, incidentally—emerge at the margins of texts on the face of it concerned with other matters, alighting somewhere before passing out of view. At times, I arrest these moments, like in Chapter 5 on Jonathan Glazer's *Under the Skin* (2013); elsewhere, in Chapter 4 on *The Shining* (novel 1977, film 1980), the insectile constitutes the largely unremarked upon but all-encompassing environment of the Overlook Hotel. More instances or encounters, not discussed at any great length, form part of the fabric of this study: textual aggregation as insect collection. In *Flights*, Tokarczuk describes a professor's enormous, verging on monstrous, knowledge of Greece, writing that it is made up of

> quotes, references, citations, painstakingly deciphered words on chipped vases, drawings not entirely intelligible, dig sites, paraphrases in later writings, ashes, correspondences and concordances. There was something inhuman in all this—to be able to fit all that knowledge in himself, the professor must have needed to perform some special biological procedure, permitting it to grow into his tissues, opening his body to it and becoming hybrid.[13]

Intimations of the post- or inhuman, iterations of the insectile: texts enter the skin, opening up bodies to a becoming-other. The writing, in this book, is similarly rhizomatic, a metamorphosing tissue-growth or, in Deleuze and Guattari's words, a little machine involving inside and outwith, alert to the ways in which the insectile swarms and is assembled within literary and cultural texts. The introduction locates the insectile in Lacanian thought and explores the 'aporia' between Lacan and Deleuze and Guattari to rethink the notion of the so-called human subject. Roughly divided up into two thematic parts, the introduction covers form and formlessness in the first part (beginning with the subsection '*Informe* and Ideal-I') and ideas of fascination in the second part ('Scenes of Fascination: II'). The material to be considered in the remaining chapters tracks the insectile—its concerns with form and formlessness as elements of fascination with either one or both of those operations—across the archives of psychoanalysis, 17th-century painting, novels from the 19th century to the present, and post-1970s film. The readings—a mode of interpretation that combines close reading, the interiority of texts, with an associative, if you wish parasitic, imagination—seek to draw out the insectile in its in-forming role in the production and deconstruction of the subject.

Informe and Ideal-I

The Return of the Native famously begins with a curious perspective, which might be traced back to a furze-cutter mentioned in the opening pages or to a similarly marginal presence—Diggory Venn, the reddleman, an extinct figure linked to the dodo—who barely registers on the 'face' of the heath, an enigmatic space sending out its own elusive gaze. Anna Feuerstein argues that both Venn and Clym Yeobright 'most full embody Hardy's ecological ethic,' moving beyond the bounded form of the so-called human subject to encompass more-than-human perspectives.[14] These emerge in the function of the subject as insect, in terms of being 'lowered' from 'the vertical to the horizontal,' to borrow Yve-Alain Bois's words, relinquishing the hierarchy (and sight) of the so-called human.[15]

In Hardy's novel, this operation, proceeding from Clym's 'morbid sensitivity to light,' takes the form of a 'daily life' of a 'microscopic sort':

> his whole world being limited to a circuit of a few feet from his person. His familiars were creeping and winged things, and they seemed to enroll him in their band. Bees hummed around his ears with an intimate air, and tugged at the heath and furze-flowers at his side in such numbers as to weigh them down to the sod. The strange amber-coloured butterflies which Egdon produced, and which were never seen elsewhere, quivered in the breath of his lips, alighted upon his bowed back, and sported with the glittering point of his hook as he flourished it up and down. Tribes of emerald-green grasshoppers leaped over his feet, falling awkwardly on their backs, heads, or hips, like unskilful acrobats, as chance might rule; or engaged themselves in noisy flirtations under the fern-fronds with silent ones of homely hue. Huge flies, ignorant of larders and wire-netting, and quite in a savage state, buzzed about him without knowing that he was a man.[16]

Rather than morbid, Clym's sensitivity to light prompts a responsiveness to, and intimacy with, 'creeping and winged things,' reconfigured as 'familiars.' Their little lives affect his own, the circumference of his activities comparable to that of a parasite, *Tineola bisselliella*, the common clothes moth 'fret[ting] a garment'.[17] Even though 'man' is still at the centre of these observations—the circuit emanates from there—the phallic subject is clearly dissipating, as if degrading, morphing into other ways of being in the world.

Clym's responsiveness and attention, prompted by blindness, are linked to the dismantling of ideological systems of verticality, that is, the normatively organised subject and its (ableist, speciesist) culture. A disability and animal studies perspective might recognise Clym's shift in terms of a 'valuing of otherness' to be traced back to the deconstruction of the phallic; his involvement with creatures he now considers his familiars indicative of a differently aligned subjectivity, immersed in an interspecies relationality.[18] Carla Hustak and Natasha Myers call this new dynamic, between an involved and distributed subject and its encounters with others, be they animal or plant (their work is on the constellation between plant and insect, subject and other), an 'affective' and 'effusive' ecology, in which there is 'never a passive or empty space between bodies'.[19] While I take issue with, and am surprised by, Hustak and Myers's coding of passivity—despite their politics, passivity appears as if in a patriarchal framework—they are effectively describing a milieu of fascination without naming it so: an affective ecology is an 'ecology of mimetic becomings,'[20] in other words, of the 'opaque, empty opening' of the force of fascination.[21] In Hardy's novel, the empty, passive space—used in

Blanchot's sense, as space of potentiality—is vibrantly alive, Clym's immediate surroundings an 'effusive ecology' of colour and noise, the latter a phenomenon without boundaries, an 'aggregate' without form.[22] Michel Serres explains that noise is not even a phenomenon, considering that the term implies an emergence out of something else, is a 'thing that appears,' according to the OED. Serres's approach resonates with the hum described in the cited passage above: eternal, constant, it is usually unremarked upon and, in its 'no-thing-ness,' is constitutive of the 'ground of our being'.[23] The horizontality or horizontalisation of the subject is indexed to the insectile here, a dimension that is concurrently microscopic and enormous, exceeding and dissolving 'man's' form.

The vocabulary I deploy (dissipation, diminishing, degrading, etc.) gives the impression of an exclusively negative operation, but its announcement or promise of an undoing is to be understood as an 'affirmative exigency,' in Derrida's words,[24] motivated by the possibility of justice and hospitality (also used in the Derridean sense). The phallic subject is a thing to be urgently undone—one origin point of such deconstructions is canonical psychoanalysis—and what I propose here is a further attempt to keep undoing it: again, still, relentlessly. My study is, as such, an anti-fascist intervention, bearing in mind the correspondences between fascist and phallic symbolisations and subjectivities, the fetishization of *das Aufrechte*, the upright and rigid, to stand up and stand erect. In the documentary film *Kleine Germanen* (2019), the German far-right publisher Götz Kubitschek and his wife Ellen Kositza speak about raising their kids to be *aufrecht*, referring at once to the spine and to an unbending, iron will: to face the other without accommodation, without welcome.[25] The verticality of the fascist imagination structures itself according to hallucinated but no less systematically upheld oppositions between form and *Unform*, proper and improper, sovereign subject and abject other, what Georges Bataille refers to as the *informe*: that which 'does not, in any sense whatever, possess rights, and everywhere gets crushed like a spider or an earthworm.'[26] Patrick Crowley and Paul Hegarty distinguish between the formless or *informe* and formlessness, noting that the latter is a state, something that lies forever outside of form, whereas the *informe* or formless is a process.[27] I do not share their view and will not reproduce their interpretation of formlessness as ontological condition. The conceptualisation of formlessness, like its adjective and as already mentioned above, is determined through specific sociohistorical and political coordinates that do not remain unchanged. Formlessness, *informe*, *Unform* are not eternal but contingent, and I use these terms indistinguishably. Though neither of the life 'forms' Bataille mentions is categorized as an insect, their existence—creatures not bestowed with any value—can nonetheless be approached according to what I call the insectile.

I derive the word and operations of the insectile from Derrida's discussion on the problematics of the subject, invariably held in suspension between inverted commas in 'Eating Well.' In this interview, he proposes to 'rearrange' the subject so that it 'no longer dominates from the centre' and suggests describing it as 'subjectile' to record the *jet* of *différance*, the dehiscence that exists at its heart.[28] Jean-Luc Nancy, with whom Derrida is in conversation in this piece, likens the subject's 'intrinsic dislocation' to a 'murmur,' the background noise of the voice of the other which prevents the subject from ever being the 'author' of its own manifestation.[29] The subject, as such, appears as 'point of passage,' on the way to taking its place elsewhere, in the locus of the other, to use Lacanian terminology: it is on the other's stage that the subject is constituted in the first instance.[30] The notation of the subjectile, the -ile affixed to the subject, inspires the deployment of the insectile as something that similarly detaches itself from the noun or concept of an insect. An affix, after all, is an appendage that is mobile, drifts, gestures in other directions, across or beyond the base form of the word, the 'form' of an insect, to instead or concurrently encompass certain 'events' associated with insects: swarming, hatching, buzzing, squirming, and so on.

This is a project about morphology, specifically about subject formation and entomological fascination, in which the insectile is thought of as different kinds of metaphorical enactment of functions linked to insects or to dreams of insects. The materiality of insect life—this denomination is, of course, shamefully inadequate, considering the huge, and at the same time dangerously falling, variety of such lives—occasionally emerges, and disappears, throughout this study, not least in relation to the process of metamorphosis, which the anatomist Sir William Harvey (1578–1657) perceived as 'the sudden crystallisation of form from formlessness'.[31] Harvey's interpretation has since been rethought and refined—at any rate, form and formlessness lie in the eye of the beholder—but his reading of the *informe*, the impure and imperfect and of its 'sudden' eruption into form (from his perspective not a process but an event) resonates with this book's concerns. The 'event' of form—changes happen inside, are hidden by the insects' external exoskeleton—occurs in Benji's 'fake time,' that incomplete temporality where we see only 'the results when the old cuticle [of the exoskeleton] is abruptly shed'.[32] Nicole A. Jacobs has written on that same temporality, 'people time,' and the ensuing limited perspectives regarding bees and the concept of labour in Shakespeare's *The Tempest*; Janina Wellmann, concerned with the pictorial representations of metamorphosis from the 17th to the 19th century, argues that the drawings served to try to understand the inner changes, metamorphosis as an epigenetic and gradual process, extending deep beyond 'fake time'.[33]

On a branch, floating in an empty, passive space, the metamorphic stages of a butterfly take place, all at the same time. Maria Sibylla Merian's illustrations, in *Metamorphosis Insectorum Surinamensium* (1706), though not the focus of Wellmann's article, are frequently unreal, despite their precision, providing impossible vantage points: a blossoming orange tree also already bears fruit; a thin branch too fragile, too unmoored, to hold the orange a caterpillar is approaching. Merian's drawings attempt to capture insect development, whose very obscenity seems reproduced in these impossible dimensions: materialising a life cycle against a blank, obscure background into which it threatens to be reabsorbed. Egg, caterpillar, pupa, butterfly, an insect's lifecycle is

> punctuated by molts, during which a new cuticle is formed and the old one shed, a process termed ecdysis. [...] The span from one ecdysis to the next is termed an instar. In most insects, there is a characteristic number of instars between hatching and metamorphosis, but the decision to begin metamorphosis typically depends on reaching a characteristic species-specific size. Therefore, suboptimal nutrition or injuries or disease might result in the intercalation of additional molts until this size is achieved.[34]

An insectile temporality, then—an instar time: layered, palimpsestic, remaindered; a time of intercalation—has to do with moulting, degradation, invagination, developments that hatch 'entomo-oneirically' in 'our' imagination. Steven Connor mentions 'entomo-oneirism' in his book on skin, in relation to fantasies of new corporeal emergences: the skin that splits and is discarded for a different body—more resistant, refreshed, enhanced—to erupt into form.[35]

My investigation, concerning an engagement with a range of formal features pertaining to cultural representations of insectility, does not assign or privilege form but is interested in (fascinated by) the moulting of forms. The insectile's formal features, as such, refer to the *informe* as much as to a fetishization of form, what I refer to as total form, derived from Lacan and the German writer and entomologist Ernst Jünger (1895–1998): a fantasy-subject of absolute density and coherence, epitomised in the coleopteran's hard exoskeleton. The insectile, consequently, pushes in both directions, the dissipation of the subject and the deliquescence of form as well as its entomo-oneiric resurgence, the terrifying 'magnificence' of a subjectivity rendered as unassailable, tight, armoured. In Lacanian terminology, this latter manifestation is an Ideal-I, the 'imago' called into (fictional) being during the mirror stage. Imago, of course, is a biological term designating the last stage of an insect's metamorphosis, in which it attains its final living form. The Ideal-I is the subject's specular appearance, the moment of its 'jubilant assumption' of the total form of its body as imago, the projection and destination of the I as image or (entomo-oneiric) dream.[36] In Lacan's theory of the mirror stage, the insectile resonates, erupting in his descriptions of the subject's transformation or, rather, inauguration:

> For the total form of his [sic] body, by which the subject anticipates the maturation of his power in a mirage, is given to him only as a gestalt, that is, in an exteriority in which, to be sure, this form is more constitutive than constituted, but in which, above all, it appears to him as the contour of his stature that freezes it and in a symmetry that reverses it, in opposition to the turbulent movements with which the subject feels he animates it.[37]

The 'maturated' subject is a fantasy form; 'more constitutive than constituted,' that form is largely aspirational, still to be achieved though it is also already over there, in the mirrored reflection. That which looks back, the total form of the subject, returns a *Gestalt*, exteriority, frozen stature, in other words an exoskeleton standing in opposition to the fragmented thing—haunting the subject in dreams—that animates the subjective form. The imago of the subject bestows the latter—which, prior to this moment, had not existed—with 'an "orthopaedic" form of its totality,' an exterior and therefore exoskeletal system that hardens into the 'armour of an alienating identity'. For Lacan, the 'fantastic anatomy' of the subject, repressing the reality of the I as *informe*, as an aggregate of disintegrating limbs and organs, can be conceptualised as fortified camp or stadium;[38] the other shape that suggests itself is the carapace of an insect body.

As much as the insectile occurs with reference to Derrida's work and its ethical possibilities, it is, then, also inscribed in relation to Lacan, whom Derrida invokes in 'Eating Well' to draw attention to his contribution to the deconstruction of the subject. The armoured *Gestalt* emerging in the mirror stage—clamping into a place elsewhere, in a remote and removed location from the fragmented body dreaming of totality—destabilises the subject's foundations at the very moment it comes into being. Broadly speaking, these Lacanian inscriptions of insectility involve the following elements:

- the mirror stage, inaugurating/'maturating' the subject whose dream-form is coleopteran and whose 'presence' is always already located in the dimension of the other. This place, as Lacan writes in his seminar on anxiety, 'represents the absence where we stand';[39]
- *objet petit a*, the remainder or residue of the subject's relation to the other; something cleaved off in the process of subjectification (*ce qu'on n'a plus*, that which the subject no longer has, with the *n'a* as the 'offcut, as something missing'), which functions in terms of a 'dialectic of cut and lack'.[40]

While I have laid out the correspondence between Ideal-I and the insectile in terms of 'ideas of enshellment'[41]—though there is more to do on this

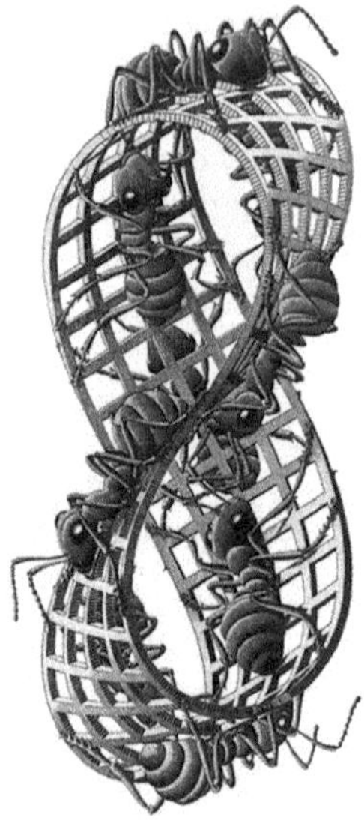

Figure 1.1 M.C. Escher, *Möbius Strip II*: woodcut of Möbius strip with ants.

note, not least with respect to the phenomena of mimicry and fascination—I have yet to speak of the insectile as 'our' other, that is, of the I bearing the insectile trace of that other. These two divergent implications (form and formlessness) of the insectile happen simultaneously in Lacan's analysis of subject formation. As the subject acquires its imago, it is at the same time 'cut' by the relationship to the other, in many ways marked as insectile in Lacanian thought. The insectile appears, most conspicuously, in *Seminar X: Anxiety*, whose 2014 edition published by Polity Press reproduces M.C. Escher's *Möbius Strip II* (1963)—a woodcut depicting ants crawling along a grid-like, uncannily turning band—on its cover (Figure 1.1).

Uncanny Turns

The Möbius strip, a continuous surface, twists inside into outside and vice versa. The twist is not a point or a 'rim' where the outside flips into its opposite but is constitutive of the structure as a whole. For Lacan, the Möbius strip illustrates subjectivity as it assembles itself in the mirror stage: it is where the ego and the unconscious, or dream and waking life, exist on the same topological surface. In *Seminar X*, Lacan writes:

> An ant walking along one of the apparent faces [of the Möbius strip] will pass over to the other face without needing to go over the edge. In other words, the Möbius strip is a surface that has just one face and a surface with just one face cannot be turned inside out. If you turn it over, it will still be identical to itself.[42]

Bruce Fink explains that this split—between outside and inside, unconscious and ego—exceeds the structure of subjectivity, is therefore 'in excess of structure,' and defines the subject as a 'split between two forms of otherness,' involving 'the ego as other and the unconscious as the Other's desire'.[43] The I assumes its form in light of the other, which authenticates 'the place labelled I' during the mirror stage, the act compelling the I 'in a flutter of jubilant activity' to turn around to seek ratification for its image.[44] The subject's presence lies somewhere else, because validated by an Other, thereby revealing the subject's non-coincidence with itself, its non-autonomy over its 'own' subjectivity. Given that the subject forms itself in relation to the other, it is always marked by that other or, more precisely, by the other's desire which determines the constitution of the subject attempting to mould (or moult) itself into the object of the other's desire.

The Lacanian algebra designates this controlling function of the desire of the other as *objet a*, the remainder, residue or also, recalling Nancy's words, the 'murmur' disturbing the place of the I. *Objet a* is the cut, that which falls away and which characterises the function of desire; Lacan derives it from the Greek word *ágalma* which, at 'first glance,' means ornament or jewellery.[45] What it really designates in Lacanian psychoanalytic discourse, however, is 'something that is inside,' an 'essence' that remains hidden, an enigma 'entirely below the surface'.[46] It is an uncanny yet still fetishized object: a hidden thing beckons, exalting a certain shine, lubricating the gaze (or glance).[47] *Objet a*'s fetishistic function entails falling under the spell of the other, whose 'fullness' or overwhelming reality constitutes the subject's lack, that which is cut out at the moment of the subject's inauguration. If the insectile arises at the level of the Ideal-I, the projected specular image structured as coleopteran fiction, it also introduces itself as that which signifies the elusive cause of desire, *objet a*. This process of signification takes root in Lacan's *Seminar X*, framed, as it is, by Escher's woodcut (Figure 1.1). The Möbius strip is one of the early topological models Lacan introduces to conceptualise subjectivity, as we have seen, as well as his notion of *extimité*, a specifically spatial understanding of the uncanny: the inside outwith, the outside within. The seminar begins not with ants though, as the frontispiece might lead us to think, but with a praying mantis to indicate the essential relationship between subject and other:

> I pictured myself [having donned an animal mask] faced with another animal [...], taken to be gigantic for the sake of the story, a praying

> mantis. Since I didn't know which mask I was wearing, you can easily imagine that I had for some reason not to feel reassured in the event that, by chance, this mask might have been just what it took to lead my partner into some error as to my identity. The whole thing was well underscored by the fact that, as I confessed, I couldn't see my own image in the enigmatic mirror of the insect's ocular globe.[48]

The praying mantis is itself a reference to Roger Caillois, informing Lacan's work on the advent of the subject.[49] The praying mantis serves to illustrate the other's radical difference, its voracious desire, which the I has to face not knowing what mask it wears, what object it is. The other's gigantic size and colossal appetite articulate its absolute presence, all the while preventing any access to its reality. It remains powerfully unknown, its demand unfathomable in the eyes of the subject-to-be, who nonetheless has to figure out its form as a response to the other's 'opaque weight'.[50] The relationship between subject and other or, rather, of the I to the other's desire, is fundamentally structured by (castration) anxiety, captured in the metaphor of the praying mantis, whose ocular globes reflect no indication of the nature of its desire. Here, the famous question *Che vuoi?*—'what does [the other] want concerning this place of the ego'—strikes the I [*je*], casting [*jeter*] it into anxiety: the *jet* of anxiety troubles the very assumption of the *je*.[51]

For Lacan, the psychodrama of subject formation plays itself out entomologically: the other is (metaphorically) insectile, the enigma of its desire shaping the object that I anxiously form myself into. The I is caught—as it needs to be—in the other's field of vision, an arrest which incarnates subjectivity, a form forever bearing the residue of the other. Something always escapes when it comes to the recognition of 'our own' subjective form and yet it is precisely this escaping object, unknown and incommunicable, that dominates the subject: this is the function of the *a*, determining the metamorphosis of the I.[52] *Objet a* 'slides in' from the outside and although Lacan calls it 'un-imaged'—it has no specular dimension and 'cannot be marked out'—it is nonetheless 'figured' as insectile: the ants on Escher's Möbius strip burrow their way into the text.[53] Even though Lacan utilises the ants to visualise the ego, rather than *objet a*, as topographical structure, there are intimations that conjoin *objet a* and insectility. An insect wanders along the surface of the ego, '[believing] from one moment to the next that there is a face [it] hasn't explored, the face that is always on the back of the face along which [it] is walking'. Exploring one face then the other without noticing a moment of cross-over, the ant, Lacan writes, is missing a 'little piece,' the element of *objet a*, which 'forms the reality of the world the insect is walking about in'.[54] This 'little missing portion' short-circuits the insect from one point to the same point on the 'other' side of the strip, but this 'scrap' or 'waste product' can't be pinned down even as it slices through the subject.[55] In other words, the link between insect and Möbius strip develops in relation to *a*, the radical lack that twists the subject into the direction of the other and which is the precondition for the subject's coming into being.

In *Seminar X*, Lacan talks about the five forms *objet a* takes, moving along a circular arrow from oral to anal, phallic, scopic and superego stages. These forms result in different orientations of the subject towards the other's desire. Accordingly, I propose the following insectile forms of *objet a*, bringing formlessness and form, outside and inside, into correlation:

- as remainder or 'waste product,' something incommunicable, *a* is *informe*; it resists signification and yet yields the incarnation of the signifier, the (fantasy) form of the subject. The elusive *objet a* prompts the subject-to-be to constitute itself into an object which shines, a coleopteran Ideal-I. *Objet a* is the formless cause of the coleopteran fiction, the total form, of the subject;
- *objet a* comes from 'an outside' that stands 'prior to a certain internalization' and, once inside, it introduces the 'distinction between ego and non-ego': it is the outside within.[56] There are powerful affective dimensions to this *extimité*, explored, and frequently negatively so, in the cultural imaginary: the delusional parasitosis in William Friedkin's *Bug* (2006), for example; the drug-related hallucinations in Donna Tartt's *The Little Friend* (2002), in which Danny Ratliff, a meth addict, thinks everything—skin, food, lungs, eyeballs, heart—is infested with insects;[57] the minutiae of 'invasion' evident at the level of language in Jonathan Lethem's *Motherless Brooklyn* (2000), in which Lionel Essrog, suffering from Tourette's, tics on his own name, 'Essrog, Essrog, Essrog, I chanted, like a cricket trapped in the wall'.[58] There is an archive of the insectile in relation to the 'irreducible *incognito*' of the subject, so that ants wandering along the surface of the ego also denote the trace of the insectile or parasitic 'in-habituated' other in the subjective make-up;[59]
- desire 'cuts' the subject in two ways: as a falling away and as a function that 'goes off to hook on wherever it can'.[60] The metaphor Lacan uses to demonstrate the falling away of *objet a* once more attaches to, or detaches from, an insect's body—exemplified by the black-beetle or the cockroach—with darts, a 'claw, a hooking object'.[61] The function of *a* as such at once grafts and cuts, hooks on and falls away. Desire, as Lacan argues, binds to the function of the cut as much as to the function of the remainder.

In Escher's woodcut, the ants, though they appear as if they could be replaced by a finger, say, tracing along the Möbius strip to demonstrate its uncanny logic, comment on the various facets of the insectile as it manifests itself in relation to the formation and in-habituation of the subject, defined by the absent presence of *objet a*. The ants might well, as Robin Purves suggests, represent the 'frantic bustle of desire as it rushes forward but always ends up tracing the same path over and over,' an interpretation alert to the ways in which insects are figured in terms not only of anxiety—the 'anti-orgasm' of profound shock; an 'insect-type swarming' threatening to collapse the nervous system—but also of desire.[62] The structuring of anxiety and desire enmesh, at any rate, as Lacan argues, not least concerning their configurations in the cultural imagination. We might refer to Alison Bechdel's *Fun Home* (2006) in this instance, her memoir about growing up and coming out, in which hormonal fluctuations coincide with and are externalised as a plague of locusts, spending years underground in a nymphal condition before emerging as winged adults. Bechdel writes that 'by the end of the first week in June,' the locusts 'had settled down to an orgy' all around her, as if their 'vibrating chorus' had '[shaken] loose the screws on some collective libidinal impulse, unleashing it into the atmosphere'.[63] Later on, watching the Watergate scandal unfold on television—whose truth is 'worming its way, like a larval cicada, toward daylight'—she is trying to understand her familial situation, obscured by her dad's secret homosexuality. She notes that 'some crucial part of the structure seemed to be missing, like in dreams I have [...] where termites had eaten through all the floor joints'.[64] These metaphors and literalisations of the larval behaviour of secrets and libidinal impulses—the word 'impulse' harbours the *informe*; a pulse is a beat that punctures, incites irruptions—reveal the insectile to be one of the central enunciations to image the 'un-imaged,' that is, the function of desire. At the same time, it shapes the fantasy state of the subject, tickled (by ants, locusts, termites) on its 'two' sides or faces, ego and non-ego, both of which are bearing the 'cut' of the other.

Figurations

Insectile subjectile—the subject implicated in the insectile—is focalised in Lacanian thinking, whose model of the Möbius strip serves to illustrate not only the inside-outside topography of the subject but also the twisting movements of the insectile. These movements have to do with insides and outsides, *Gestalt* or form and the *informe*, while they similarly apply to the project's engagement with the posthuman. Part of what this book is doing is relocating or returning the 'event' of the posthuman inside the subject and to offer the insectile as multiple, frequently conflicting, functions of articulation or disarticulation (form and formlessness) of that subject. It should by now be clear that by subject, I understand precisely that unstable proposition that Derrida discusses in 'Eating Well,' emphasising the internal displacement (the *jet*) the *sujet* already carries within itself. I'm by no means the first to think of posthumanism's 'turn' from the outside, as prostheses or appendages, to the inside,[65] not even as it pertains to insects which, according to Jussi Parikka, have replaced the already nostalgic cyborg as prototype of the posthuman. In his book *Insect Media* (2010), Parikka thinks of media technologies as bestial, of how figurations (a word to which I'll come back) of insects are 'transposed' into various technologies as 'carriers of intensities'.[66] One of the main aspects of Parikka's analysis is a thinking beyond the body and the signifier to focus on affects, to think about assemblages rather than discrete entities. Assemblages, of course, as well as affect theory, have accrued a rich philosophical tradition in recent years and can themselves be 'transposed' into the discourse of the *informe* or, to refer to Sebastian Vehlken's work, the continually changing parameters of formations, deformations, formattings and transformations.

More so than Parikka, I look to Vehlken's *Zootechnologien* (2012) to provide another rung to this project's scaffolding because of its focus on form. Vehlken is interested in the 'figuration' of the swarm, itself a word that must be put under pressure. A figure, after all, has a distinct form, while a figuration is the process of moving toward, consolidating into, form. Rosi Braidotti employs it slightly differently in *Metamorphoses* (2002), writing that a figuration renders 'a decentred and multi-layered vision of the subject as a dynamic and changing entity,' as a kind of 'living map,'[67] a sense I invoke and at the same time want to keep at bay. It is, rather, the suspension—between form and formlessness; figure and the unfigurable—that I seek to highlight in using the term 'figuration,' hanging between citation marks and in-between states of realisation and derealisation. Rather than its vitality, I wish to emphasise its oscillation or slippage between states, the word signalling a flickering or shimmering, moving in and out of focus, away and toward figure and form. There is another meaning to the term, as Steven Connor reminds us while writing on the 'figure' of the fly. *Figura* refers to both form and face, continuing that the fly and insects more generally are of uncertain form, not least because they seem to lack a face. 'The fly,' Connor writes, 'is always itself, but that self does not form a figure, nor yet exactly figure a form. More than anything else, the fly is a figure for this unfigurability'.[68] This aspect of 'not forming a figure, nor yet exactly figuring a form' is crucial with respect to the 'figuration' of the insectile: it registers the mechanics of passing into and out of form.

In his book, Vehlken is alert to the ways in which swarms are made to mean and are integrated into, as well as produced as, forms of knowledge [*Formen des Wissens*]. He investigates how the 'figure' of the swarm changes over time, from calling up an 'epistemic horror,' its appearance something that can't become form, to a rendition of admirable, aspirational life.[69] (We might recall *Alien* [1979] here or also *Starship Troopers* [1997], both of which feature organisms of epistemic horror and, at the same time, total perfection. In Paul Verhoeven's *Starship Troopers*, the

'Arkellian sand beetle' has no ego and is therefore the 'perfect selfless member of society' seamlessly integrated into the functioning body of the superego; in *Alien*, the cyborg Ash admires the 'purity' of the creature which, 'unclouded by conscience, remorse or the delusions of morality,' is pure drive.[70]) Vehklen's study traces a genealogy of the swarm's emergence, circa 2000, as media-technological manifestation of an optimised and totally controlled multiplicity. His investigation begins around 1900, with a chapter called 'Deformations,' and the start of an ordering of the swarm in the context of mass psychology. Between 1900 and 1930, ethologists, including a great number of amateur researchers, developed their own model of analysis that tried to look at the phenomenon of the swarm outside of a clearly anthropomorphic agenda, that is, to understand group psychology at hand of swarm behaviour. The overarching concern, in this chapter about 'Formations,' is a framework or field of visibility in which the swarm is not only 'object,' however difficult to define—indeed, it keeps evading analysis despite or because of the new techniques used to determine its properties—but a network of functions. The 'object' really is understood as mode of operation, giving and receiving feedback from its *Umwelt*, a perspective most prominently argued by Jakob von Uexküll, who writes that 'an animal is an [...] event,' not a 'thing' but an occurrence.[71] The swarm's object-hood, around 1980, returns into the focal point with sonar technology and research into schools of fish, so Vehlken continues in a chapter on 'Formatting'. The purposes of this particular method are to eliminate interferences, the background noise of the sea, and to thereby isolate the swarm as 'smooth' space, a space that is regulated, existing in opposition to the surrounding chaos. Paradoxically, these acoustically generated computer images render the swarm, 'a body without surface' according to Leonardo da Vinci,[72] into pure surface: the 'body' dissipates to become space, topology. Even though the aim is to establish form, this practice returns the swarm's *Ungestalt*, its morphology that of a blob which, in the final chapter of Vehlken's book, 'Transformations,' is instrumentalized as an 'organism' that behaves optimally, directed by its own internal logic, thus arising as archetype, an ideal appearance, of form.

The varying means by which to discover or bestow form that Vehlken discusses in *Zootechnologien* make one thing clear: the instability of form and figurations, apt to destabilise and 'degrade' into formlessness, whose viscosity, noise or entropic operations are similarly temporal, to be 'rescued' or reshaped as triumphant model of form. The insectile, expressive, as it is, of this continual shift between form and *informe*, is not, then, per se indicative of a radical politics, and as such distinguishes itself from Deleuze and Guattari's becoming-insect, nonetheless exerting its considerable influence on the current project. This is not to say that the processes of becoming invariably evade reterritorialization, which Deleuze and Guattari occasionally intimate in *A Thousand Plateaus*. Rather, I want to emphasise its deployment in contemporary theoretical thought. Whenever becoming is invoked—in accordance to the movements it overwhelmingly performs in *A Thousand Plateaus*, that is, as anti-molar force, attempting to decimate the molar power of the phallic subject and its 'dreadful Oedpial atmosphere'[73]—it maps a politics of difference, oriented toward the outside, uprooting 'me' from 'my' kind and subjecthood. Such interpretations are compelling, responding to Deleuze and Guattari's mapping of the processes of becoming as they are for the most part proposed in *A Thousand Plateaus*, but they tend to pass over those instances in which becoming restores form. At first sight, this might seem like a careless assertion, inattentive to the general temperament of Deleuze and Guattari's work, but what it does, instead, is engage with those moments where becoming slips, or is in danger of slipping, into restoration.

One such point is the Body without Organs (BwO), the dismantling of the self which, through its disarticulation, can enter into communication with other intensities, planes, and so on and which, in contrast to becoming, is treated more cautiously in *A Thousand Plateaus*. A BwO is unformed matter, Deleuze and Guattari write, but all the same, it is subject to what they call strata, 'acts of capture' that are form-giving.[74] These strata 'spawn their own BwO's [sic], totalitarian and fascist BwO's,'[75] demanding a vigilance directed at those apparently destratified assemblages masking their configuration as despotic organisations. BwOs can, consequently, be 'botched,' played out so as to yield fascist, cancerous things, a transversality—from form-defying and resistant to form-generative or preserving—that has to affect becoming too, especially considering that the BwO is articulated as a 'becoming-machine of the organism,' as Daniel Smith shows.[76] Every organ has its function, is in fact pure function and thus is not recognisable as form: it is circuit, it circulates intensities; it is kinematic movement and energy transformation. A little machine to 'be plugged into other collective machines,' the BwO behaves as an operation of becoming, acting against strata, the organisation of the organism, the fantasy of the subject.[77] And yet it is weighed upon by the 'judgment of God,' intent on stratifying the BwO. Hence, Deleuze and Guattari advise us to be alert to the production of false apparitions like the paranoid body or the drugged body, both of which are 'emptied and dreary' manifestations, as well as the 'cancerous' BwO, the 'BwO of the State, army, factory, city, Party'.[78] The latter is the main reason for their caution, heedful of 'each instant, each second' a BwO turns cancerous, its becoming falling into stratification, the organisation of form (a tumour) and the 'stifling body of subjectification'.[79]

I would like to maintain this element of watchful observance when it comes to discussing becoming and becoming-insect more specifically and propose a sacrilegious act of interpretation or conceptualisation. If the 'insectile' at once records that shift away and towards form, the instant, each second, becoming might coagulate and triumphantly deliver up a form, the figure that suggests itself is the Möbius strip. For Deleuze and Guattari, Lacan, along with 'General' Freud, is a lurking 'priest'. Both Lacan and Freud are representatives of the 'judgement of God,' seeing desire in terms of lack, a 'negative law' incompatible with the delirious force of desiring machines put forward in *A Thousand Plateaus*.[80] And yet I propose the Möbius strip as an illustration demonstrating the mobilisation of the insectile and its involutions of outside and inside, in this case revolving around form and formlessness. Franklin, in 'The Context of Forms,' reminds us that the phenomena of form and formlessness need to be examined together, arguing that processes of formalisation—like the enclosures of land; the objectification of living bodies—constitute the *modus operandi* of racial capitalist production.[81] Formlessness is, as such, the precondition for form and, to further recall Vehlken's study on the changing mobilisation of the swarm, products of socio-historical and political conditions that value form over the *informe*. The Möbius strip stands for those uncanny turns where form always already references its apparent 'other' face, and formlessness stratifies, can be captured, commodified—as form not yet discovered—and recruited as part of the logic of formalisation. The one is apt to pass into or appear as the other at any time or point in space.

Missed Encounters

To put Deleuze and Guattari into relation with Lacan tends to break the rules of engagement set up between these thinkers. The hegemonic position insists especially on Deleuze and Lacan's incompatibility, the former critical of Lacanian psychoanalysis, 'inscribing in desire the negative law of lack,' making Lacan into the 'priest' of that lack and negative law.[82] In their book on the 'disjunctive synthesis' between Deleuze and Lacan, Boštjan Nedoh and Andreja Zevnik note that contemporary debates about the two are structured around this irreducibility and are, further, prevented by prejudice.[83] Once both thinkers are drawn on at the same time, responses to their deployment usually either are profoundly uneasy, mindful of and abiding by the general consensus or, marking consensus as law, refuse and forbid their encounter altogether. Consensus, as such, dictates practices of reading prior to acts of interpretation, and consensus is also what is sought as a result of these processes of reading and interpretation: it is both precondition and desired outcome. Given that it apparently can't, in this case, be achieved, any discussion must consequently fail, because unable to proceed from and/or arrive at that consensus, often reason enough to prevent discussions from occurring in the first place.

Yet other protocols of reading can be set out, taking their cue, perhaps, from Derrida's 'missed' encounter with Hans-Georg Gadamer at the Goethe Institute in Paris in April 1981. The symposium, which ran on the subject of 'Text and Interpretation,' was an attempt to bring hermeneutics and deconstruction into dialogue with each other and was generally thought to have been a disappointing event. Its failure was largely ascribed to Derrida, whose paper did not refer to Gadamer's work, nor did it address the questions Gadamer had asked in his talk. Derrida's questions to Gadamer, in turn, seemed tangential to the latter's argument, so that the symposium appeared to demonstrate the impossibility of dialogue between the two positions. As Colin Davis argues, however, what it showed is the difficulty, not the impossibility, of an encounter, which should not be judged on the basis of failing to achieve consensus but pursued precisely because it might not do so. A 'failed' encounter leaves behind 'an active, provoking trace,' in Derrida's words, 'thereby promising more future than would have been the case with a harmonious, consensual dialogue'.[84] The symposium, in light also of Derrida's later reflection on it in *Béliers* (2003), effectively expounds his reading practice, proceeding in terms of what he calls the countersignature, which attempts to engage with and preserve, rather than assimilate and deny, the absolute singularity of the other:

> There is, as it were, a duel of singularities [between the reader and the 'event' of the text], a duel of writing and reading, in the course of which a countersignature comes both to confirm, repeat and respect the signature of the other, of the 'original' work, and to lead it off elsewhere, so running the risk of betraying it, having to betray it in a certain way so as to respect it, through the invention of another signature just as singular.[85]

The symposium, putting into play Derrida's countersignature, was all the more successful for having failed, for forcing that trace, the remainder, to be discovered, even experienced as something that maintains the impenetrability of the other, whom it runs the risk of betraying. The much more serious act of betrayal, from Derrida's perspective, would have consisted in settling for a consensus, an indication not of respect or hospitality but of disavowal: of the other's signature, of one's own countersignature and, as such, of one's responsibilities before that other, whose singularity can be confirmed only without compromise. That which prevents consensus, the irreconcilable moments between a text and its reader, between positions that can't be reduced to some common denominator eliminating difference, might finally allow for encounters whose 'difficulty' is not to be resolved but to expose oneself to.

An encounter between Deleuze, Guattari and Lacan shapes this investigation, passing by way of Derrida's 'law' of reading, which includes taking account 'of what is shared (in the sense of both participation and division, of continuity and the cut of separation)'.[86] Peter Klepec, without mentioning Derrida (who nonetheless seems to 'preside' over his essay), imagines one such other encounter between Deleuze and Lacan, and argues that the defining aspect of any encounter, even the 'happy' ones, is the element of surprise or shock, the mystery that surrounds each *point de rencontre*.[87] Klepec shows that both Lacan and Deleuze view the encounter as an event that takes place unexpectedly, arrives from elsewhere and cannot be scripted in advance. These are 'points' that are shared and at the same time enable us to think that which cuts, like, for example, the function of the unconscious. In the introduction to their edited collection, of which Klepec's essay forms part, Nedoh and Zevnik speak in those terms, that is, of the problematics of the encounter, of an enigmatic occurrence defining the relationship between Lacan and Deleuze:

> Perhaps the two thinkers offer an esoteric debate: one of responses and reservations without naming names; one in which the reader can be often in doubt whether Deleuze and Lacan really talk to each other when they are talking to each other and, simultaneously, a conviction that they talk to each other precisely and only when they do not name each other.[88]

A number of possible other encounters, departing from the consensus of their impossible dialogue, have been considered beyond Nedoh and Zevnik's collection; Geneviève Sartor uses Lacan's theory of the sinthôme to demonstrate his engagement with *Anti-Oedipus*, and Janell Watson, in her book on Guattari (always considered the 'minor' writer compared with Deleuze), analyses his 'cosmic-scale ontological drawing[s]' to illuminate the extent to which Guattari remaps or re-models Lacanian thought.[89] In *The Neutral* (2002), Roland Barthes is queried on evoking Lacan and Deleuze on the subject of desire by one of his students, who asks Barthes on how he proposes to resolve the conflict between the two. Barthes responds by observing that the neutral, or the desire for the neutral, is aporetic, does not seek a resolution nor the repose of a conclusion: Lacan in conversation with Deleuze means thinking the aporetic without undoing or eliminating it.[90] This book follows suit, by thinking the contingency of an improbable encounter and, above all, by looking at the cut, at that which cuts, an endeavour cutting with Derrida's practice of reading and writing. I proposed insectile forms of *objet a* above, all of which cut into the subject: the cut is the 'structural operator' of subject formation, giving rise to its *Gestalt*, the image-form and exoskeletal site of subject coherence, as well as to its internal split.[91] It sustains yet further relations to the insectile in terms of the latter's etymological origin which, rather than point of stability, behaves as an open system of connections and associations in Derrida's 'Fourmis,' functioning like a little machine driven by elements cutting in, 'marcotting' themselves to the word 'ant'.[92]

Fourmis

In 'Fourmis,' an appendix—which itself has something to do with cutting: it is added but its attachment is not essential—to a book of interviews conducted between Hélène Cixous and Mireille Calle-Gruber, Derrida reflects on the word 'ant,' which arrives from one of Cixous's dreams recounted to Derrida on the telephone. Without being able to anticipate its effect, which he admits to noticing only at the point of writing, this ant, Derrida notes, 'make[s] its way within [him], insinuating itself between experiences that resemble song as much as work, like the animals of the fable'.[93] It is a busy contribution, ostensibly about sexual difference, but reading it generates an impression of being dragged off through 'thousands of meanings' on the back of this ant; the essay further teems with errors, typos and errant commas, accidentals in an already noisy piece.[94] I mentioned squirming earlier, referring to insectile (and technologized) life that, for the most part, is hidden from view, teems in the cracks, in (telephoned) dreams and the unconscious, in instar time, beneath surfaces or stones or skin. A 'figure' of tiny multitude, the insectile swarms in texts and media more generally, insinuating itself into them and into 'you.' Derrida reminds us that the word insect comes from *inseco*:

> which means to cut, to dissect, at times to tear with the teeth (*dentibus aliquid insecare*), to put into small pieces. The frequentative *insector* means 'to pursue without respite,' to be on the heels, to hasten energetically, to seduce, perhaps to court, to harass, to go after—etc. As '*insecta*,' this sort of genus [*genre*], of quasi-genus specified by thousands of species, the ant is a cut invertebrate (the word means *cut*, it names the cutting), that is to say divided into small strangulations by so many annuli.[95]

'Fourmis' performs these manoeuvres of insinuation, dissection, multiplicity, harried pursuit, seduction, deferral (not least through the 'fly-specks' of punctuation marks, the cuts of the comma),[96] intervention (what is an intervention if not a cut, an incision into the text) and parasitic asides. Language might be a virus, but it is also 'swarm-entity,' as Connor writes in relation to Beckett's *The Unnameable* (1953).[97] The insectile signifies rupture, being interrupted, distracted, and stands in relation to an outside breaking in or an outside situated right at the 'heart' of subjects, laws, the so-called human 'genre'. In 'The Law of Genre,' Derrida speaks about the essential corruption, an internal division, impurity and deformation affecting genres, not simply literary form—intent on trapping writers like 'little bug[s],' according to Eileen Myles—but also 'membership in sets' more broadly.[98] Genre in general means form, designed to contain or detain, but each form 'invaginates' a 'principle of contamination, a law of impurity, a parasitical economy,' cutting across concepts of genre, manifestations of forms.[99] 'Fourmis' contributes to this debate about genre, drawing out that which is already intimated in 'The Law of Genre,' where the insectile appears in the emphasis on the parasitical. We might also glimpse it in Derrida's analysis of the non-closure of genres through a trait that, like a fly—a 'formless form'—enters to participate in any given codes.[100]

The insectile structures and destructures the subject's 'membership' in a genre as well as its relationship to language—further analysed in Chapter 6—and it also characterises the operations of deconstruction, which Derrida defines as parasitology, 'disrupting writing, inscription, and the coding and decoding of inscription'.[101] The ant, appearing at the beginning of 'Fourmis,' yields an 'adventure of reading and interpretation,' crawling

> with thousands of meanings, with a thousand and one images, with a thousand and one sexes, it cuts itself in the middle (four/mis), it can lose its two wings or only one (because the ant, Hélène's insect, is a winged insect [*à aile*], an insect that is classed among the winged insects, the hymenoptera), it is put, and it's put in the oven, the he-ant put in motion/the oven turned on [*le fourmi en marcke* [sic]], in the little oven and the great oven of all the incinerations, it makes, once it's been cut in two, sentences forward and backward, up to the end or halfway, it gives everything, it furnishes food and drink, the *fourme*, that is to say the form *I*, it goes in the oven or the furnace both the crust and the soft part, it's good like the bread one shares and eats in the family—and families are also anthills—but it's also something to vomit like the inedible itself.[102]

The footnote to this passage—chasing meaning 'backward and forward' through a delirium of associations—draws attention to the many ways the sentence teems with ants: '*fourmiller, mille, milieu, four, mis, mi-chemin, fournit, fournil, mie, famille, fourmillière* ...'.[103] The sense of confusion, generated by the incessant movement of images, word associations, words divided up and combined anew—a parasitic cascade—stems from an ant metamorphosing through what seems like incongruous forms. At the same time, a single ant also always invokes plurality. Despite its minuscule figure, the ant mobilises an 'innumerable multiplicity' engendered through the mode of writing, chasing a figure into the future, whose appropriate punctuation mark is, indeed, the ellipsis, suggesting infinitude, an openness to include yet more associations.[104]

As a practice of reading and writing, an 'adventure' staged through the figure of an ant, Derrida's approach resonates with Deleuze and Guattari's in *A Thousand Plateaus*, where the insectile similarly appears as principle of rupture and circulation. Specifically, they link ants to their concept of the rhizome, belonging to the multiple, the pack or swarm. A botanical term, referring to subterranean stems that travel horizontally—unlike the tree or root, giving rise to a 'cosmos' of verticality—the rhizome is mobile, assuming diverse modes of being or becoming. It establishes connections, but connections involve cuts; the rhizome agglomerates and breaks down, decentring language 'onto other dimensions and other registers':

> a rhizome may be broken, shattered at a given spot, but it will start up again on one of its old lines, or on new lines. You can never get rid of ants because they form an animal rhizome that can rebound time and again after most of it has been destroyed.[105]

While it erupts occasionally throughout *A Thousand Plateaus*—in an orchid-wasp configuration, for example, or lines of thought proceeding like 'tiny ants,' or intimations of 'ants... ants... ants...' in the conjunction 'and... and... and...' of the rhizome[106]—the insectile is analogy for, rather than metaphor of, multiple textual operations in Derrida's work, including the mechanisms of interpretation. This analogy traces throughout Derrida's writings, thus making space for the possibility of another encounter between Lacan and Deleuze and Guattari, given its insistence on that which cuts and which both informs and deforms, parasitically infests. At any rate, that possibility is alive for Lacan as much as for Deleuze and Guattari, bearing in mind that the latter acknowledge that the rhizome can find 'a foothold in formations that are Oedipal or paranoid or even worse,' adding that 'it is even possible for psychoanalysis to serve as foothold, in spite of itself'.[107] Derrida helps, as such, to think the insectile cut as a committed practice of intervention, which occurs, on the one hand, in relation to Deleuze and Guattari's process of becoming and, on the other, concerning Lacan's deployment of the Möbius strip. I cut and splice the two together in the very concept and event of the insectile, tracing becoming-insect on an uncanny structure articulating at once formlessness and the concept of form.

Scenes of Fascination: II

This introduction began with a scene of fascination between two lovers, reuniting at a village festival held outdoors after both have led unhappy marriages. The dance, where they are swept up amongst other couples, is itself a fascinating event, compulsive in its 'ceaseless glides and whirls' into which Wildeve and Eustacia are launched:

> Through the length of five-and-twenty couples they threaded their giddy way, and a new vitality entered her form. The pale ray of the evening lent a fascination to the experience. There is a certain degree and tone of light which tends to disturb the equilibrium of the senses,

> and to promote dangerously tender moods; added to movement, it drives the emotions to rankness, the reason becoming sleepy and unperceiving in inverse proportion; and this light fell now upon these two from the disk of the moon. [...] The air became quite still, the flag above the waggons which held the musicians clung to the pole, and the players appeared only in outline against the sky; except when the circular mouths of the trombone, ophicleide, and French horn gleamed out like huge eyes from the shade of their figures. The pretty dresses of the maids lost their subtler day colours and showed more or less of a misty white. Eustacia floated round and round on Wildeve's arm, her face rapt and statuesque; her soul had passed away from and forgotten her features, which were left empty and quiescent, as they always are when feeling goes beyond their register.[108]

Fascination, as the passage above demonstrates, is not a property but a complex relation that arises between, and exceeds, subject and object; it is not to be found in any single location but is instead excessive, shattering forms. Eustacia experiences it as vitality, produced by the other dancing couples, a gliding swarm, whose movements create eddies, propelling her and Wildeve into their choreography. The light plays a part, too, combining with the movement to 'disturb the equilibrium of the senses' and to 'promote dangerously tender moods,' emotions driven to 'rankness,' a noun usually deployed in relation to vegetation. In Scots, the corresponding adjective, 'rank,' continues to circulate in terms of copiousness or profusion; its contemporary usage elsewhere is limited to describing something unpleasant, especially relating to smell. It can, however, also refer to female animals in heat, a gendering that passes judgement and occasionally shifts to encompass women: 'rank daughters,' as Daniel Defoe, cited in the OED, writes in *True-Born Englishman* (1701), having 'Receiv'd all Nations with Promiscuous Lust'. The large number of obsolete or rare expressions that the OED lists for adjective and adverb, still in evidence at the time of publication of Hardy's novel, include references to great speed or force—as in a rank storm—or, as already mentioned, relate to vegetation, vigorous and luxuriant growth, an abundance haunted precisely by excess. Emotions driven to rankness and dangerously tender moods are, as such, indicative of 'extravagant' states, unrestrained manifestations of emotional investments tending towards the obscene and sending 'reason' or, in other words, the ego to sleep. The ego becomes 'unperceiving,' it drifts blindly, floats, is caught up in an exuberant, mobile rankness.

Fascination, then, is rank vitality, dangerous to the ego which loses its 'form'. Eustacia's face becomes 'rapt and statuesque,' her 'soul' passes away, forgetting her features, which are left 'empty and quiescent'. That which animates and defines her (soul, ego, face, features) is eroded, her form depleted, suspended in still air, the arresting moment of fascination.

The ambivalent scene that Hardy paints—determined by vitality and loss or absence—is more generally indicative of the phenomenon of fascination which, though prompted through ceaseless movement and a certain quality of light, is simultaneously characterised as still-standing, a frozen moment, timeless in the sense of being without or outside time, in a way eternal. In this 'empty, dead time,' as Blanchot calls fascination's curious temporality, the ego is at once addressed and dismantled. Objects, after all, are not invariably fascinating for each and every subject, responding only to those objects that somehow establish a relation to itself. (On this note, I can't resist the temptation to invoke Marco Reus, German football player and Borussia Dortmund captain, who crept his way into one of my articles on being fascinated and the philosophy of collecting. The peer reviewer of the piece remarked that the reference to Reus was 'out of place,' considering that he 'seems hardly famous enough to merit exemplary status'. They suggested I replace the 'comparatively minor' footballer with someone 'more universally famous,' namely Cristiano Ronaldo, a suggestion which went unheeded, not only because it misses the point. Fascination is not property of an object but exists between subject and object, with which the former identifies excessively or, I should say, rankly.) That relation between subject and fascinating object remains an enigmatic one, highly seductive and blinding, resembling, according to Oliver Harris, the 'unattainable' Lacanian Real of desire.[109] The 'thing' that fascinates can only be represented negatively, so Harris continues: it is a 'kind of nothingness' and resists symbolisation, repeatedly calling the subject back for more, to seek not meaning but, rather, the 'experience of failing to see and know properly'.[110] The event of fascination consequently commands a surrender on the part of the subject, looking yet 'unperceiving,' held captive by an 'image' compelling identification: the ego is the form hailed by fascination even as it sends it to sleep.

In *Seminar II* (1954–1955), on the formation of the ego, Lacan writes:

> Fascination is absolutely essential to the phenomenon of the constitution of the ego. The uncoordinated, incoherent diversity of the primitive fragmentation gains its unity in so far as it is fascinated. Reflection is also fascination, jamming.[111]

The mirror stage is milieu of fascination as well as the *initium* of desire, that cut marking the subject at the moment that it hallucinates itself as imago and subsequently enters the symbolic order. A fragmented, formless 'thing' finds its articulation as unity only through the function of fascination. Held in the gaze of the other, the pre-subjective assemblage—the 'body in pieces,' a 'subject who is no one'—assumes *Gestalt*, a transformation into a finished or total form which corresponds to the reflection it receives from that other.[112] Lacan talks of little machines in this instance, an incomplete little machine structuring itself in relation to another machine having already perfected its unity, so that the 'movement of each machine is [...] conditioned by the perception of a certain stage by another'.[113] He argues that the ego totally depends—it does not exist prior to this moment—on the element of fascination, an argument guided by Caillois's work on mimicry. In his article, 'Mimicry and Legendary Psychasthenia,' Caillois considers the adaptation of certain insects to their environments: the Phyllia or leaf insects, some species of butterflies and praying mantises, whose legs 'simulate petals or are curved into corollas,' thereby inserting themselves into the plant world.[114] He proposes that the mechanism that drives this process of assimilation is fascination and that mimicry, rather than a defensive or protective formation against a predator, is a 'luxury,' the imitating mechanism induced by the 'magical hold' of the space the insect finds itself in.[115] The governing principle of the insect's adaptation to space consequently is that of excessive identification. Caillois talks about a 'mutual organisation' or 'reciprocal topography' between insect and space, even of 'teleplasty,' according to which the mimicking insect reproduces, on its own body, the form and relief of the fascinating object/space.[116] He extrapolates this phenomenon, which he calls 'temptation by space,' to schizophrenia or legendary psychasthenia, a disturbance of the relations between so-called human subject and spatiality, the latter percolating into the former to the point that it disappears the ego.[117] The fascinating object/space enthrals the subject, which is compelled to arrange itself so as to converge with the object: the I no longer knows where to find itself.

Caillois's observations are to be found throughout Lacan's writings, informing his analysis of the gaze of the other and the function of desire, the reduction of the subject to zero in the scopic field of the other, as he writes in *Four Fundamental Concepts* (1973).[118] For Lacan, the function of fascination yields form, specifically the form of the subject at the moment of the mirror stage: fascination is that which gives rise to the I in the image, as a response to the gaze, of the other. He develops this argument in 'The Mirror Stage' as well as in his seminar on the ego in Freud's theory; the essential function of fascination is further apparent in *Seminar X*, edging in at the limits of the mirror. The I is fascinated by the contents of the mirror, the illusory aspect of the image, but there's something it can't see, lying beyond the mirror, past its edges: the provenance and location of the gaze.[119] Lacan mentions the 'other' side of fascination in *Seminar X*, gesturing once more to Caillois and his mimicking insects. That which fascinates is the function of the gaze which, though formative, is also the point 'where all subjective subsistence seems to get lost, to be absorbed, and to leave the world behind'.[120] The species example which Caillois gives and Lacan calls into evidence is *Smerinthus ocellata*, the eyed hawk moth, whose eye-shaped spots are vehicle of fascination arresting the 'other party': the eye's mystery is 'point of irradiation,' where the subject is captured in the grid of desire.[121] There is surrender here and absorption, the form of the phallus apparently triumphing, recalling Freud's 'Medusa's Head' (1922). Freud describes the mythological encounter (another scene of fascination) of a boy-child with Medusa, whose sight makes the male spectator 'stiff with terror,' turning him to stone.[122] The same process of stiffening happens in Freud's case study of the Wolfman, who reacts in 'phobic terror' to the beating of a butterfly's wings, 'not so very far,' as Lacan remarks in *Four Fundamental Concepts*, 'from the beating of causation, of the primal stripe marking his being' as subject caught in that gird of desire.[123] Wolfman's dream of the tree with its perched, immobile wolves holding the dreamer in their gaze turns the child to stone,

> paralysed by this fascination to the point that one may conceive of what gazes back at him in the scene, and which is invisible on account of being everywhere, as nothing but the transposition of the arrested state of his own body, here transformed into a tree. [...] The subject is no more than an erection in this grip [of the gaze] that makes him a phallus, that freezes him from head to toe, that arborifies him.[124]

Wolfman's dream is itself mirror stage or belongs to the same order of subject 'erection'. Fascination signals the 'zero point' when the subject (or pre-subjective assemblage) takes its place, once more or for the first time, in the field of the other.[125] The phenomenon is, as such, linked to formlessness—it is the scene where 'subjective subsistence' is lost—and to phallic form, which it returns and installs. The very notation of the I, raised up, asserts the presence of this arborified subject, nonetheless always troubled by the 'primal stripe' or stain that escapes its vision and which is everywhere and nowhere at the same time. The subject seeks to announce its presence by becoming stiff—the I projects and hallucinates itself as phallic—all the while being subject to a vision that, because it fascinates, holds or floats the viewer in a state of captivation.

In *The Space of Literature* (1955), Blanchot writes that the event of fascination exposes the subject to the 'intimacy of the outside,' the 'indecisive' event producing the outside as inside and vice versa.[126] In this state, the I sinks 'into the neutrality of a featureless third person,' so that the fixity characteristic of the fascinating scene is not tied to form but to the *informe*; the recurring verb is to sink, or to be drawn, into this indeterminate milieu.[127] The gaze of the other, 'incessant and interminable,' he continues, is a 'depthless deep,' a 'light one sinks into,' where 'objects sink away' and 'collapse into their image,' which

> instead of alluding to some particular feature, becomes an allusion to the featureless, and instead of a form drawn upon absence, becomes the formless presence of this absence, the opaque, empty opening onto that which is when there is no more world, where there is no world yet.[128]

Blanchot's writing, affectively resonant, describes a mirror without naming it so; a mirror, after all, is a deep without depth. It provides coagulated light to sink into and a 'space' where objects collapse into images in order to show that fascination, a vision without the ability to see, suspends 'the power to give sense,' to sense the world, to sense the self.[129] He is, as it were, arresting fascination as element that transforms the I (back) into no one, so as to interpret it as radical possibility: the subject of language surrendering to the outside. Blanchot's insistence on the fundamental passivity of the fascinated subject supplements Lacan's work on the phenomenon, which nonetheless and simultaneously, as we have seen, dissolves the subject and pushes it toward total form. Not interested in form, not even forms that draw on absences—he is concerned with writing, and words are forms drawing on things that are missing—Blanchot evokes mirrors to disappear into their abyss without being concerned with images or ego-structures as armoured totalities or coleopteran fictions. For those perspectives, we need Lacan: coleopteran fictions form the basis of his analysis of subject constitution, whose dream of an armoured totality always threatens to catch up with the reality of its subjective 'form'. Lacan's work reminds us that the metamorphoses that occur in the milieu of fascination are never final or assured—there is no permanent or almighty phallus[130]—but haunted by their other manifestations and apparent opposites as well as by the eternal gaze of the other, a praying mantis integrating the subject into relations with the insectile.

I want to bring this introduction to a close by offering a not entirely coherent reading—at any rate, there are always remainders—of a collage by Max Ernst, which is attentive to the constellation of elements that I have sought to prove above and carry through the rest of this study: namely the connections between un/forms of subjectivity, the insectile and fascination. In Ernst's *Une Semaine de Bonté* (1934), a 'novel' composed of surrealist collages derived from wood and steel plate engravings of French popular fiction—often revolving around *crimes passionnels*—a collage integrated into a section titled 'Jeudi' shows a figure in a bourgeois interior looking into a small hand-held mirror (Figure 1.2). The figure, like all those included in this particular section of the novel, has the face of an Easter Island stone head and is dressed in a robe, standing at a commode on which lie, face down, another small, oval mirror and a fine-toothed comb. Diagonally below the mirror held by the figure, in which a shadow is reflected, a praying mantis is engaged with its prey, a fly, perhaps, grasped in its raptorial legs. In a window to the back of the figure, a nude lurks, her eyes largely obscured by the window's wood frame. The other thing of note is a strange illustration inserted into an ornamental frame, depicting clusters of cells like those of a virus, blood vessels or *Drosophilidae* eggs, floating against a featureless white background. Ernst's work has, of course, been linked to Freud (whom he had read) and to psychoanalysis more generally. André Breton, for example,

Figure 1.2 Max Ernst, collage from *Une Semaine de Bonté*: collage, on the left-hand side, of Easter Island statue looking into hand-held mirror; in the background, a partially obscured nude; a praying mantis is located diagonally below the mirror.

Max Ernst, Ile de Pâques, from *Une Semaine de Bonté* (1934). Ernst Max Ile de Pacques © ADAGP, Paris and DACS, London 2022. Max Ernst, 'Volume V: Le Rire du coq, L'Ile de Pâques, L'Intérieur de la vue, La Clé des chants,' from *Une Semaine de Bonté* (1933–1934, published 1934). Volume V from a five volume serial novel with 182 line blocks after collages, page (each): 10 5/8 × 8 1/16″ (27 × 20.5 cm). Publisher: Éditions Jeanne Bucher, Paris. Printer: Georges Duval, Paris. Edition: 812. The Louis E. Stern Collection. Acc. no.: 828.1964.E. © 2022. Digital image, The Museum of Modern Art, New York/Scala, Florence

compares *La femme 100 têtes*, a similar 'novel' from 1929, to a 'meticulous reconstruction of a crime witnessed in a dream,' and Matthew Beaumont argues that Ernst's phantasmagoric *Une Semaine de Bonté*, with its 'violent interventions' into 19th-century pictography, 'portray[s] nothing less than the unconscious' of that century and culture.[131] For Breton, who organised an exhibition of Ernst's collages in Paris in May 1921, the first encounter with them constitutes 'a kind of originary moment, [...] surrealism's primal scene,' as Rosalind Krauss notes, who further analyses the intrusions of truncated body parts into the 'banal solidity' of bourgeois settings by pointing toward Lacan.[132] Although she focuses especially on the disembodied hands entering Ernst's *La femme 100 têtes*, always appearing to beckon and to welcome the spectator-visitor into these interiors, the connections to Lacan arise elsewhere in Ernst's work too as well as in relation to other intrusive part-objects.[133] Part-objects, indicative, as they are, of the gaze, designate the subject's central lack; what the collages reveal is the phenomenon of castration, that is, the subject as fragmented assemblage, as well as the event of fascination.

Seen in this light, of the rupture or of intrusions (the very process of collage), the mirror stage scene at the start of 'Jeudi' initiates a story about voyeurism and sexual assault, the witnessing of primal scenes, recurring in Ernst's work, while suggestive of the fundamentals of subject formation in Lacanian thought. Accordingly, 'Jeudi' gestures towards a number of aspects Lacan puts forward as defining the relation between subject and other. To begin with, the collage might well correspond to the fable that Lacan relates in *Seminar X*, of him 'donning' a mask with which to face the praying mantis/the other. There are more cuts to bring up: the subject fascinated by the contents of the mirror and turning to stone, erecting itself as a phallic form in response to the mirrored reflection it sees; the other existing beyond the limits of the mirror, the praying mantis as representation of the gaze, the outside, a remnant that constitutes the subject taking its place in the grid of desire. Then there is the lurking nude, at first more Freudian than Lacanian, usually the object of the gaze and of aggression, whose sight is obscured and thereby yields a similar blindness in the spectator, as Beaumont argues concerning another collage in *Une Semaine de Bonté*.[134] Adapting Beaumont's interpretation to this illustration, the question raises itself as to what position we occupy in this setting: outside, looking in, without the ability to see (are we fascinated or 'castrated' by what we witness), or inside being looked at. In Lacanian terms, what is our place in this loop of desire, sending our gaze outwards and reeling it back in—*fort, da*—leaving it unmoored, unable to achieve satisfaction. The circuit of desire involves seeing, being seen, or more precisely being hailed, and/or the inability to see; the gaze being sent out and returning or refusing to be met. A lost, enigmatic object always escapes our inscription in this loop, the irrepressible movement of the libido which Lacan, in *Four Fundamental Concepts*, calls an 'organ,' like clusters of cells, the 'membranes of [an] egg' flying off, 'something extra flat' that 'moves like an amoeba'.[135] Or, perhaps, this organ moves like ants running around the Möbius strip of the subject, enunciated as so-called human in the field of insectile otherness.

The Shape of Things to Come

The introduction, or Chapter 1, has set out the methodological framework for this book on the insectile subjectile, that subjective 'form' inhabituated by the insectile. The theoretical debates and figures mentioned here circulate through each of the chapters below. Even though the chapters are divided into two sections, titled 'Form' and '*Informe*,' the book is alert to the ways in which these two discourses are 'formatted,' are either culturally cathected or abjected. The deconstruction of the so-called human underlies both sections of the book, demonstrating the insectile as that which enunciates the so-called human but whose politics differs depending on how it is imagined.

Chapter 2, initiating the book's focus on form, engages with the fantasy of the insect body arising in relation to Ernst Jünger's work who, more so than Kafka, charts the *Verwandlung* I want to map out here. The argument is intent not on observing alienation, then, as in Kafka's novella, but on witnessing the assemblage of a subject fortified, armoured, assuming an Ideal-I as coleoptera. There is a fascist politics at work in this imagination, which I discuss in Jünger's work, putting forward a *Typus* of 'man.' *Typus* arises out of entomological practice and refers to systems of classification; it is paragon and fantasy of form. I further place Jünger in relation to Madeleine Dewald and Oliver Lammert's rhizomatic 2002 documentary film *Vom Hirschkäfer zum Hakenkreuz*, gathering an associative chain between stag beetle [*Hirschkäfer*] and swastika [*Hakenkreuz*]. Across these texts, the insect is produced as fascinating and fetish object, vehicle of fascist desire, articulating the exoskeleton of an aspired subjective form.

Chapter 3 continues the section on form, which is set up, in H.P. Lovecraft's writings, as that which threatens to dematerialise and must be preserved at all costs. The chapter investigates the phonic materiality of sound, specifically of buzzing voices in Lovecraft's 1930 short story 'The Whisperer in the Darkness.' The insectile is configured as a trope for the outside and as a formless entity, the latter rendered as an enfleshed voice, theorised through Jean-Luc Nancy's *À l'Écoute* (2002), Mladen Dólar, Douglas Kahn and Alexander G. Weheliye among others. I am concerned with the interplay between form and formlessness as it pertains to sound, bearing in mind the production of form and formlessness as political categories, not ontological givens. I argue that what we see play out in Lovecraft's short story, as well as in his oeuvre more generally, is a sustained valorisation of form and thereby position myself against recent scholarship, notably Graham Harman's *Weird Realism* (2012), claiming Lovecraft as a writer offering a deconstruction of 'man' through perspectives other than 'human.' The latter, as I will show, remains absolutely understood, in Lovecraft's work, according to what Sylvia Wynter calls the 'coloniality of Being.'

Chapter 4 analyses *The Shining*, both Stephen King's novel (1977) and Stanley Kubrick's 1980 film adaptation. I investigate a becoming-insect that occurs with respect to Jack Torrance as a result of his enmeshing with the being of the Overlook Hotel, configured as wasp-force across novel and film. I deviate from Deleuze and Guattari's understanding of becoming in order to think through how becoming or swarming comes to function as politics of control. The first part of the chapter maps the processes of becoming onto *The Shining* and proceeds to investigate the phenomenon of fascination at work in novel and film by invoking Caillois's work on mimetic insects. The final part focuses on the insectile sound-image of the Hotel, above all using Michel Chion's writings. I use *The Shining* in order to think through and interrupt the automatic understanding of becoming-insect/swarming as operating beyond the sovereignty of the father as well as to reflect on *The Shining*'s continued parasitic apparitions in contemporary culture and politics. This chapter, as such, proposes an addition to Deleuze and Guattari's discussion of becoming in *A Thousand Plateaus* by offering up the 'memories' of an assemblage repressed in their work: memories of a Hotel-Daddy-Wasp-Machine.

Chapter 5 launches the section on the *informe* by analysing Jonathan Glazer's 2013 film *Under the Skin*. It stages a close reading of the film as it pertains to the insectile or, in other words, to an entomological imagination structured according to apparent opposites, form and formlessness. On the one hand, the insectile is instrumentalised as racialised technology centred on the face, rendering a fixity of form, and, on the other, it is coded as that which undoes precisely this logic of form. In a first instance, I am paying attention to the faciality—seeking recourse to Deleuze and Guattari's writing on this note—of the unnamed, alien woman (Scarlett Johansson), constructed as insectile. The second part of the chapter lends an ear to the film's sonic environment, the buzz of its extra-diegetic score, once more using Chion but also Greg Hainge, Michel Serres and Jennifer Barker. I further respond to Sheryl Vint's claim that the film cannot offer an ethics of difference, suggesting a position—the insectile subjectile—from which an 'improper' ethics of difference might begin to be thought.

Chapter 6 is concerned with writing, or language more generally, and the insectile; the extent to which writing establishes a scene of fascination linked to insects; as a process (including that of speech and of reading) that bears the traces of an entomological fascination and which, as a result, has the potential to generate the insectile subjectile. The chapter is, in a first instance, indexed to Maurice Blanchot's work on fascination and the movement of writing in *The Space of Literature* (1955) while further turning to consider *The Infinite Conversation* (1969), in which Blanchot speaks of writing as a 'relation of the third kind'. In the texts considered here, writing functions as space of fascination pushing into relations of the third kind: the chapter begins with Vladimir Nabokov's *Ada or Ardor* (1969) and proceeds through several other encounters with insects—A.S. Byatt's 'Morpho Eugenia' (1992), Jonathan Lethem's *Motherless Brooklyn* (2000), etc.—before arriving at Clarice Lispector's 1964 novel *The Passion According to G.H.* In each of these texts, the enigma of fascination implicates the subject in the insectile, and the scene of writing taking place constitutes itself as a scene of intimacy—or extimacy, using Lacan's term—with the insect other that fundamentally disrupts the being of the subject.

Chapter 7 revolves around the occurrence of flies in 17th- and 18th-century Dutch and Flemish still lifes, approached, also, through Bruce

Chatwin's 1988 novel *Utz*. It is interested in demonstrating that still lifes function as extinction event for the phallic subject, held captive in the pure exteriority of the other's gaze. A crucial aspect of this argument is the concept of fascination and the function of the gaze developed in Lacanian thought, which organises my reading. Lacan's analysis of Hans Holbein's *The Ambassadors* (1533), though a figurative painting, serves as a further reference point for still life paintings in which the 'form' of the fly occurs, using Steven Connor's various studies on flies to reflect on the insects' 'unfigurability.' In a first instance, then, I analyse how the mechanism of fascination works and how its various relations produce planes of desubjectification. The chapter then moves on to consider the fly as *trompe l'oeil* and parergonal element—Derrida's concept for that which arrives from elsewhere and disturbs the relationship between artwork and 'remainder'—to demonstrate that the interiority of still lifes gives way to the 'outside,' a space that destructures the so-called human subject. The chapter proposes that the historical form of the still life, as a polemic against the so-called 'human' subject, gains new critical importance in the age of species extinction.

The focus of the coda is Freud's *Wolfman* (1918) in order to think about the racialising gaze, rendered as white wolf. The overarching trope of this chapter is the cocoon, articulating the psychic reality of the racialised subject, which experiences itself as insectile or larval. The Wolfman, who is Russian and in therapy with a Jewish psychoanalyst, is not at first sight 'epidermalised,' to use Frantz Fanon's word, but nonetheless ranked according to what Alfred J. López calls 'hierarchies of whiteness,' applied to both analyst and analysand. The chapter thus reads Freud's case study of the Wolfman to think about the racialised rendition of the I as insectile. While it looks at *Wolfman* in its specifically (interwar) Viennese context, it also to a certain extent breaks with that context to think about the later Nazi and neo-Nazi mythologisation of the wolf and, more generally, the terrorizing presence of the white phallic subject.

Notes

1 Thomas Hardy, *The Return of the Native* (Cambridge, UK: Cambridge University Press, 2021), 344–345.
2 Hardy, *The Return of the Native*, 217.
3 Hardy, *The Return of the Native*, 218.
4 Hardy, *The Return of the Native*, 238.
5 Jacques Lacan, *The Seminar of Jacques Lacan, Book X: Anxiety*, ed. Jacques-Alain Miller, trans. A. R. Price (Cambridge & Malden: Polity Press, 2015), 241.
6 Olga Tokarczuk, *Flights*, trans. Jennifer Croft (New York: Riverhead Books, 2018), 25.
7 Vilém Flusser, *Vampyroteuthis Infernalis*, trans. Valentine A. Pakis (Minneapolis & London: University of Minnesota Press, 2012), 5.
8 Flusser, *Vampyroteuthis Infernalis*, 42.

9 Lauren Berlant, *The Queen of America Goes to Washington City: Essays on Sex and Citizenship* (Durham & London: Duke University Press, 1997), 18.
10 Colson Whitehead, *Sag Harbor* (London: Vintage Books, 2011), 91 & 153.
11 Seb Franklin, 'The Context of Forms,' *World Picture*, Vol. 11, Summer 2016, 1–23.
12 Dagmar Herzog, 'Desire's Politics: Félix Guattari and the Renewal of the Psychoanalytic Left,' *Psychoanalysis and History*, vol. 18, no. 1 (2016), 29 & 11.
13 Tokarczuk, *Flights*, 376.
14 Anna Feuerstein, 'Seeing Animals on Egdon Heath: The Democratic Impulse of Thomas Hardy's *The Return of the Native*,' *19: Interdisciplinary Studies in the Long Nineteenth Century*, Vol. 26, 2018, 4.
15 Yve-Alain Bois, 'The Use Value of "Formless",' in *Formless: A User's Guide*, ed. Yve-Alain Bois & Rosalind E. Krauss (New York: Zone Books, 1997), 26.
16 Hardy, *The Return of the Native*, 317 & 322.
17 Hardy, *The Return of the Native*, 354.
18 Sunaura Taylor, *Beasts of Burden: Animal and Disability Liberation* (New York & London: The New Press, 2016), 60. I also make reference here to Carla Hustak and Natasha Myers's article 'Involutionary Momentum: Affective Ecologies and the Sciences of Plant/Insect Encounters,' *differences*, vol. 23, no. 3 (2012), 74–118.
19 Hustak & Myers, 'Involutionary Momentum,' 78, 103 & 105.
20 Hustak & Myers, 'Involutionary Momentum,' 93.
21 Maurice Blanchot, *The Space of Literature*, trans. Ann Smock (Lincoln & London: University of Nebraska Press, 1989), 33.
22 Michel Serres, *Genesis*, trans. Geneviève James & James Nielson (Ann Arbor: University of Michigan Press, 1999), 2.
23 Serres, *Genesis*, 13.
24 Jacques Derrida, *Points ...: Interviews, 1974–1994*, ed. Elizabeth Weber, trans. Peggy Kamuf & others (Stanford: Stanford University Press, 1995), 83.
25 Mohammad Farokhmanesh & Frank Geiger (dir.), *Kleine Germanen: Eine Kindheit in der Rechten Szene* (Little Dream Pictures GmbH, 2019).
26 Georges Bataille, *Encyclopaedia Acephalica: Comprising the Critical Dictionary & Related Texts*, trans. Iain White (London: Atlas Press, 1995), 51.
27 Patrick Crowley & Paul Hegarty, 'Introduction,' in *Formless: Ways in and Out of Form*, ed. Crowley; Hegarty (Oxford, etc.: Peter Lang, 2005), 12.
28 Derrida, *Points ...*, 268 & 260.
29 Derrida, *Points ...*, 265 & 259.
30 Derrida, *Points ...*, 259.
31 Brian W. Ogilvie, 'Order of Insects,' in *The Life Sciences in Early Modern Philosophy*, ed. Ohad Nachtomy (Cary: Oxford University Press, 2013), 228.
32 James W. Truman, 'The Evolution of Insect Metamorphosis,' *Current Biology*, vol. 29, no. 3 (2019), R1242.
33 Nicole A. Jacobs, *Bees in Early Modern Transatlantic Literature: Sovereign Colony* (New York: Routledge, 2021), 29–53; Janina Wellmann, 'Die Metamorphose der Bilder,' *NMT: International Journal of History & Ethics of Natural Sciences, Technology and Medicine*, vol. 16, no. 2 (2008), 183–211.
34 Truman, 'The Evolution of Insect Metamorphosis,' R1253.
35 Steven Connor, *The Book of Skin* (London: Reaktion, 2004), 178–179.
36 Jacques Lacan, 'The Mirror Stage as Formative of the Function of the I as Revealed in Psychoanalytic Experience,' in *Écrits: A Selection*, trans. Alan Sheridan (New York & London: Routledge, 2001), 2.
37 Lacan, 'Mirror Stage,' 2.

38 Lacan, 'Mirror Stage,' 4.
39 Lacan, *Anxiety*, 47.
40 Lacan, *Anxiety*, 117 & 123.
41 Rachel Murray, *The Modernist Exoskeleton: Insects, War, Literary Form* (Edinburgh: Edinburgh University Press, 2020), 13.
42 Lacan, *Anxiety*, 96.
43 Bruce Fink, *The Lacanian Subject: Between Language and Jouissance* (Princeton: Princeton University Press, 1995), 46.
44 Lacan, *Anxiety*, 41 & 32 and 'Mirror Stage,' 76.
45 Jacques Lacan, *Transference: The Seminar of Jacques Lacan, Book VIII*, ed. Jacques-Alain Miller, trans. Bruce Fink (Cambridge & Malden: Polity, 2015), 167.
46 Lacan, *Transference*, 138 & 162.
47 Sigmund Freud, 'On Fetishism,' in *The Complete Psychological Works of Sigmund Freud*, Vol. XXI, trans. & ed. James Strachey (London: Hogarth & the Institute of Psychoanalysis, 1950), 198.
48 Lacan, *Anxiety*, 5–6.
49 The white, male, etc. subject, needless to say; the praying mantis recurs, of course, in misogynistic discourses (we might think of the femme fatale in Film Noir), an aspect that I push, however, aside and bury in these footnotes.
50 Lacan, *Anxiety*, 61.
51 Lacan, *Anxiety*, 6 & 7.
52 Lacan, *Anxiety*, 88.
53 Lacan, *Anxiety*, 102 & 60.
54 Lacan, *Anxiety*, 136.
55 Lacan, *Anxiety*, 136 & 174.
56 Lacan, *Anxiety*, 102.
57 Donna Tartt, *The Little Friend* (London: Bloomsbury, 2002), 440.
58 Jonathan Lethem, *Motherless Brooklyn* (New York: Vintage, 2000), 260.
59 Lacan, *Anxiety*, 107 & 76.
60 Lacan, *Anxiety*, 102.
61 Lacan, *Anxiety*, 167.
62 Robin Purves, email correspondence, 08.12.2020. On the 'anti-orgasm,' see Anna Burns, *Milkman* (London: Faber & Faber, 2018), 79 & 295.
63 Alison Bechdel, *Fun Home: A Family Tragicomic* (London: Jonathan Cape, 2006), 156 & 157.
64 Bechdel, *Fun Home*, 172 & 173.
65 See, for example, Jacques Derrida, '*Geschlecht* II: Heidegger's Hand', in *Deconstruction and Philosophy: The Texts of Jacques Derrida*, ed. J. Sallis (Chicago & London: The University of Chicago Press, 1988), 161–196; David Wills, *Dorsality: Thinking Back through Technology and Politics* (Minneapolis & London: University of Minnesota Press, 2008); Jussi Parikka, *Insect Media: An Archaeology of Animals and Technology* (Minneapolis & London: University of Minnesota Press, 2010); Dominic Pettman, *Creaturely Love: How Desire Makes Us More and Less Than Human* (Minneapolis & London: University of Minnesota Press, 2017).
66 Jussi Parikka, *Insect Media: An Archaeology of Animals and Technology* (Minneapolis & London: University of Minnesota Press, 2010), xiii.
67 Rosi Braidotti, *Metamorphoses: Towards a Materialist Theory of Becoming* (London: Polity, 2002), 2 & 3.
68 Steven Connor, *Fly* (London: Reaktion, 2006), 31 & 32.
69 Sebastian Vehlken, *Zootechnologien: Eine Mediengeschichte der Schwarmforschung* (Zürich: diaphanes, 2012), 14. [My translation. All subsequent translations mine.]

70 Paul Verhoeven (dir.), *Starship Troopers* (Sony Pictures, 1997) & Ridley Scott (dir.), *Alien* (20th Century Fox, 1979).
71 Cited in Vehlken, 146.
72 Cited in Vehlken, 13.
73 Gilles Deleuze & Félix Guattari, *A Thousand Plateaus: Capitalism and Schizophrenia*, trans. Brian Massumi (London & New York: Continuum, 2004), 257.
74 Deleuze & Guattari, *A Thousand Plateaus*, 45.
75 Deleuze & Guattari, *A Thousand Plateaus*, 181.
76 Daniel Smith, 'What Is the Body without Organs? Machine and Organism in Deleuze and Guattari,' *Contemporary Philosophy Review*, Vol. 51, 2018, 96.
77 Deleuze & Guattari, *A Thousand Plateaus*, 178 & 179.
78 Deleuze & Guattari, *A Thousand Plateaus*, 178 & 180–181.
79 Deleuze & Guattari, *A Thousand Plateaus*, 180.
80 Deleuze & Guattari, *A Thousand Plateaus*, 171 & 19.
81 Franklin, 'The Context of Forms,' 3.
82 Deleuze & Guattari, *A Thousand Plateaus*, 171.
83 Boštjan Nedoh & Andreja Zevnik, 'Introduction: On a Disjunctive Synthesis between Lacan and Deleuze,' *Lacan and Deleuze: A Disjunctive Synthesis*, ed. Nedoh & Zevnik (Edinburgh: Edinburgh University Press, 2017), 1–2.
84 Colin Davis, *Critical Excess: Overreading in Derrida, Deleuze, Levinas, Žižek and Cavell* (Stanford: Stanford University Press, 2020), 45.
85 Jacques Derrida, *Acts of Literature*, ed. Derek Attridge (New York & London: Routledge, 1992), 69.
86 Derrida, *Acts of Literature*, 68.
87 Peter Klepec, 'For Another Deleuze–Lacan Encounter,' in *Lacan and Deleuze: A Disjunctive Synthesis*, ed. Boštjan Nedoh & Andreja Zevnik 15.
88 Nedoh & Zevnik, 'Introduction,' 5.
89 Genvième Sartor, 'A Joycean Interface: Re-Territorialising between Deleuze and Lacan,' *University of Toronto Quarterly*, vol. 87, no. 2 (Spring 2018), 1–15; Janell Watson, *Guattari's Diagrammatic Thought: Writing between Lacan and Deleuze* (London: Bloomsbury, 2009), 4.
90 Roland Barthes, *The Neutral*, trans. Rosalind E. Krauss & Dennis Hollier (New York: Columbia University Press, 2005), 68–69.
91 Lacan, *Anxiety*, 121.
92 Lacan, *Anxiety*, 143.
93 Jacques Derrida, 'Fourmis,' *Hélène Cixous Rootprints: Memories of Life & Writing*, ed. in Hélène Cixous & Mireille Calle-Gruber, trans. Eric Prenowitz (London & New York: Routledge, 1997), 119.
94 Derrida, 'Fourmis,' 119. I'm also referring to Steven Connor, 'Making Flies Mean Something,' a lecture given at the *Beckett and Animals* Conference at University of Reading on 26 September 2009. Accessible via: http://stevenconnor.com/flymean.html [Accessed 17 December 2020], 6.
95 Derrida, 'Fourmis,' 121.
96 Connor, 'Making Flies Mean Something,' 6.
97 Connor, 'Making Flies Mean Something,' 8.
98 Jacques Derrida, 'The Law of Genre,' trans. Avital Ronell, *Critical Inquiry*, vol. 7, no. 1 (Autumn 1980), 57 & 59. Eileen Myles cited in Moira Donegan, 'Gay as in Happy: On Maggie Nelson,' *n+1*, Issue 23. Available via: https://nplusonemag.com/issue-23/reviews/gay-as-in-happy/ [Accessed 22 January 2021].
99 Derrida, 'The Law of Genre,' 59.
100 Derrida, 'The Law of Genre,' 65.

101 Jacques Derrida, *Positions*, trans. Alan Bass (Chicago & London: University of Chicago Press, 1981), 95–96.
102 Derrida, 'Fourmis,' 119.
103 Derrida, 'Fourmis,' fn 2, 126.
104 Derrida, 'Fourmis,' 120.
105 Deleuze & Guattari, *A Thousand Plateaus*, 8 & 10.
106 Deleuze & Guattari, *A Thousand Plateaus*, 13, 24 & 27.
107 Deleuze & Guattari, *A Thousand Plateaus*, 16.
108 Hardy, *The Return of the Native*, 333–334.
109 Blanchot, *The Space of Literature*, 31; Oliver Harris, *Williams Burroughs and the Secret of Fascination* (Carbondale & Edwardsville: Southern Illinois University Press, 2003), 18.
110 Oliver Harris, *Williams Burroughs and the Secret of Fascination* (Carbondale & Edwardsville: Southern Illinois University Press, 2003), 18 & 17.
111 Jacques Lacan, *The Seminar of Jacques Lacan: The Ego in Freud's Theory and in the Technique of Psychoanalysis, Book II*, ed. Jacques-Alain Miller, trans. Sylvana Tomaselli (Cambridge, New York, New Rochelle, Melbourne & Sydney: Cambridge University Press, 1988), 50.
112 Lacan, *The Ego in Freud's Theory*, 54.
113 Lacan, *The Ego in Freud's Theory*, 51.
114 Roger Caillois, 'Mimicry and Legendary Psychasthenia,' trans. John Shepley, *October*, Vol. 31, 1984, 20.
115 Caillois, 'Mimicry and Legendary Psychasthenia,' 25 & 30.
116 Caillois, 'Mimicry and Legendary Psychasthenia,' 23.
117 Caillois, 'Mimicry and Legendary Psychasthenia,' 28.
118 Jacques Lacan, *Four Fundamental Concepts of Psychoanalysis*, trans. A. Sheridan (New York & London: W.W. Norton & Co., 1998), 76–77.
119 Lacan, *Anxiety*, 73.
120 Lacan, *Anxiety*, 241.
121 Lacan, *Anxiety*, 241.
122 Sigmund Freud, 'Medusa's Head,' in *Writings on Art and Literature*, trans. James Strachey (Stanford: Stanford University Press, 1997), 273.
123 Lacan, *Four Fundamental Concepts*, 76.
124 Lacan, *Anxiety*, 260.
125 Lacan, *Anxiety*, 242.
126 Blanchot, *The Space of Literature*, 31 & 25.
127 Blanchot, *The Space of Literature*, 30.
128 Blanchot, *The Space of Literature*, 32 & 33.
129 Blanchot, *The Space of Literature*, 32.
130 Lacan, *Anxiety*, 240.
131 Matthew Beaumont, 'Cutting up the Corpse: Agatha Christie, Max Ernst and Neo-Victorianism in the 1930s,' *LIT: Literature, Interpretation, Theory*, vol. 20, no. 1–2 (2009), 13 & 18.
132 Rosalind E. Krauss, *The Optical Unconscious* (Cambridge, MA: MIT Press, 1994), 42 & 81.
133 Krauss, *The Optical Unconscious*, 82.
134 Beaumont, 'Cutting Up the Corpse,' 20–21.
135 Lacan, *Four Fundamental Concepts*, 197.

2 *Homme-Insecte*

Form, *Typus*, Fetish

In this chapter, I am concerned with the form and composure, countenance or cramp of the 'human,' as it is implicated in the other, notably the insect. In the resulting formation—*homme-insecte*—the insectile, often associated with that which slips, with the trace or, in Lacanian terminology, the stain of absolute difference, is deployed as radical departure from such movements to instead operate as a dream of armouring. It accrues consistency and form, in other words, and, rather than indicative of a becoming, it renders a fantasy of being, more specifically the fantasy of the insect body. The designation *homme-insecte* derives from Joyce Cheng's article on surrealism's engagement with mimetic metamorphosis which, among others, considers Roger Caillois's essay on insects and mimicry, which he published in *Minotaure* in 1935. In her article, she cites one P.E., whose poem 'Un Visage dans L'Herbe' [A Face in the Grass] appeared in the journal in 1933. The poem describes the redistribution of a face in grass: 'après l'insecte-feuille, l'homme-feuille' [after the leaf-insect, leaf-man].[1] The unknown face launches an investigation into figures like Caillois's mimicking insects or Salvador Dalí's 'êtres-objects' that, according to Cheng, 'serve as means for the surrealist circle to recuperate forms of passivity'.[2] Cheng thus demonstrates the link between the phenomenon of metamorphosis and passivity, corresponding to a process of becoming, substantially and involuntarily rearranging the subjective makeup: becoming is not a decision the I can take. By contrast, the way it appears here, *homme-insecte* is the figuration of an ideal hatching out of a fascist and fascinated imagination. My focus is, first and foremost, the German writer and entomologist Ernst Jünger (1895–1998), whose considerable oeuvre—composed of novels, essays, short stories, diaries and memoirs—illustrates the *Verwandlung* I want to map out. Jünger remains one of the most controversial figures in German (and German-speaking) culture, despite, or because of, the numerous awards he won (including the *Große Bundesverdienstkreuz mit Stern* in 1977 and the Goethe Preis in 1982); the sheer volume of his writings, more often than not concerned with the mythical image of the 'human,' alone seems to

DOI: 10.4324/9781003322757-2

consolidate just such a figure, articulated, across these writings obsessed with masculinity and hardened carapaces, as *homme-insecte*.

Rather than Kafka, it is Jünger, then, who constitutes the main concern of this chapter, even if it still converges on what Cristopher Hollingsworth calls the 'self as insect,' a topos pioneered by Kafka to represent the 'crushing effect' of an unloving and regimented world on the subject, its regression in the face of a 'crisis of belonging'.[3] The argument below, however, is intent not on observing alienation but, rather, on witnessing the assemblage of a subject fortified, armoured, assuming an ideal-I as coleoptera. Lacan's term for this image or for the subject's transformation into a specular fiction during the mirror stage is imago, itself pertaining to the entomological, to the final stage of an insect's metamorphosis. This imago or *Gestalt* is a well-built figure: frozen stature in a symmetrical arrangement, Lacan speaks of it as an 'orthopaedic form,' the 'armour' of a body that, fragmented, fantasises the mantle of a total, shell-like form.[4] The subject, according to Lacan, exists discordantly in relation to its own identity, always insufficient, hallucinating its broken body-image into a vision and form of totality. The fragmented form returns in dreams, Lacan continues, although it is the totality-form—Lacan speaks of it in terms of a fortress or a stadium—that really is the dreamwork: the dream of the insect body is the fantasy of the fragmented subject wishing to be armoured, total form. Jünger's prolific writing, itself seeking to shore up authority, is central to understanding such a process of ego fortification, a discourse which has become yet more pronounced during the COVID-19 pandemic. Various groups like QAnon, anti-vaxxers and anti-maskers mobilize the conception of a powerful, invincible ego-form, a radicalised supra-individual resistant to infection; this imagination of a (virus-repellent, immunised) sovereign agent sits at the 'fortified' heart of Jünger's coleopteran Ideal-I.[5]

For Deleuze and Guattari, to map means to allow the circulations of intensities, to be open to exteriorities, principles of connection that are not necessarily evident at first sight: maps are linked to rhizomes, tubers, decentred dimensions.[6] To approach the subject, Jüngerian *homme-insecte*, I perform Deleuze and Guattari's practice of mapping, or of reading and writing, while further using a methodological conceit borrowed from Madeleine Dewald and Oliver Lammert's rhizomatic 2002 documentary film *Vom Hirschkäfer zum Hakenkreuz*, gathering an associative chain between stag beetle [*Hirschkäfer*] and swastika [*Hakenkreuz*]. This chain is assembled by a machine, the *Historionaut*, evoking a space/time explorer and archivist, trawling through databases to show a dimension where the stag beetle is rooted to Nazi and, more generally, fascist ideology. Insects put into relations with fascism—including from the perspective of satirical anti-fascist critique or fear of fascism 'among us, [...] in the heart of England'—are more broadly in the air in the 1930s; this particular entanglement appears, for example, in Virginia Woolf's 1938 book *Three Guineas*, in which the fascist dictator is figured as insect, at first 'curled up like a caterpillar on a leaf,' threatening to develop into imago.[7] John Heartfield's *Deutsche Naturgeschichte Metamorphose*, a photomontage published in *AIZ* (*Arbeiter-Illustrierte Zeitung*) in 1934, similarly depicts the three stages of development into fascist imago in the history of the Weimar Republic: Friedrich Ebert, the first president of the Republic, begins the caterpillar stage; Paul Hindenburg, the second *Reichspräsident*, who proclaimed Hitler *Reichskanzler* on 30 January 1933, hangs in a cocoon at the chrysalis stage; finally Hitler emerges in the shape of a Death's-Head hawkmoth whose abdomen is emblazoned with a luminous swastika.

The *Historionaut* is a framing device within the film itself—as well as conscious-making memory machine, remembering forgotten associations—which dispenses the need for linearity or a conventional narrative logic instead arranging material allusively. The connections between the images in the film often have to be conjectured, subjected to a reading akin to an interpretation of dreams. The result is a generative work, whose 'anti-form' in and of itself already protests the rigid formality of its subject matter, that is, the structuration of desire according to fascist thought. The *Historionaut*—we could call it a folding machine: it folds one text into another—is endogenous to this chapter, setting up a rhizomatic string of associations, an assemblage composed, often latently, through the psychoanalytic writings of Freud and Lacan. *Historionaut* behaves 'psychonautically' or, to give a salute to Alexander Trocci, as cosmonautic explorer-machine of inner space, tracing a *typus*.[8] Jünger's term for the form of a subject that exceeds the manifestation of an individual entity, *typus* arises, as Dan Gorenstein explains, out of entomological practice. It is a designation that, at the beginning of the 20th century, was used to 'refer to individually prepared insects, specimen descriptions, recognised but marginal specimen [*Seitenstücke*] in scientific collections, or also particularly immaculate finds in private collections'.[9] *Typus* is paragon, a model or standard that is made, put into place, all the while remaining fantasy form: it is representation, ideal, 'affirmation of a certain being'[10] beyond the subject.

The rhizome is anti-genealogical, but my methodology nonetheless seeks to bring into focus a genealogy of a being that is 'entomo-oneiric,'[11] the product of an entomological fascination and fetishization, as well as fascist creature. The skin, in this dream of the insect, functions as a shell, polished, hardened. The transformation into *homme-insecte* is to be understood not as a becoming-other but, instead, as a forming of the ego as exterior armour, making reference to Wilhelm Reich. The organism's surface layer, Reich argues, is composed in response to pressures—arriving from the exterior world as well as from the libidinal forces of the unconscious—shaping the subject, whose structure, as a result of these pressures, has over the years congealed into a defensive mechanism that he calls 'character armour'. The formation and the preservation of this armour—a process of form; it determines how a subject formally behaves or reacts—establish the subject's economy, maintaining 'his'[12] (neurotic) balance. Character armour, the form mobilised against pressure, accumulates across time and is embedded in the psyche but stored in the musculature of the body: it is 'moulded expression' of psychic, narcissistic defence,[13] a permanent 'cramp' defining a particular subject-hood. The subject, seen in this light, is 'matter of muscle cramping,'[14] an embodied defensive form.

Homme-Insecte is, thus, a project about form, the fetish of form, in the sense of a particularly exalted and resplendent (adjectives that recur) *typus* of 'man'. I am, further, attentive to systems of classification—Jünger, after all, had an impressive collection of coleoptera amounting to 40,000 specimens—and the taxonomy of writing, both of which attest to processes of formalisation that similarly appear under the aegis, as it were, of the armour. The first part of the chapter ('Form') explains the territorialisation of becoming-insect through a turn toward Reich and the media and communications theorist Vilém Flusser's reading of Reich; the second ('*Typus*') is concerned with Jüngerian assemblages as armourings, particularly in his *Käfertagebücher*; the final part ('Fetish'), placing Jünger in relation to, and as secret subject of, *Vom Hirschkäfer zum Hakenkreuz*, reflects on the 'irreducible materiality'[15] of the insect body as fetish. Especially in *Subtile Jagden* (1967), Jünger's impressions of his fascination with coleoptera, he produces the insect as fetish object, vehicle of (fascist) desire, articulating the exoskeleton of an aspired subjective form.

Form

In 'Waspish Segments,' Jessica Burstein analyses the nexus of fictions surrounding the soldierly body prosthetically reworked, after World War I, into new, harder, form. Her focus is largely Wyndham Lewis, though her essay further encompasses Ezra Pound as well as Caillois and Jünger, noting that 'most of the characters in the story [she's] been telling were fascinated by insects'.[16] She continues that 'the public taste for entomology was not born with war' but that the militarisation of life 'was concurrent with the fascination for culture's carapaces, for things with "the works on the outside": French locomotives, insects, soldiers'.[17] As such, she discusses the lines of communication between the 'human' and the insect or, at any rate, a particular dream of the insect, an investigation that informs my own, standing at odds with the displacements often attributed to becoming-insect.

As Burstein demonstrates, there is a different story to be told about the propinquity between insects and 'man,' which has to do with totality and Being instead of becoming and dispersal. Nonetheless, my chapter departs from Burstein's argument in that the insect-subject, here, exists uneasily between the desire for total form and its impossibility, bearing in mind that totality is a fantasy, an imago to counteract what Lacan calls a 'dehiscence at the heart of the organism,' the trauma of incompleteness, of premature birth.[18] The armour is designed to radically deny that trauma as well as the subject's ties to the other, coded as feminine and, in the context of insects, as multiple, in terms of the 'liquidity' and formlessness of the swarm. *Homme-insecte*, consequently, in this piece about order, about how things are placed into systems, also has to be understood in terms of a critical relationship between form and formlessness, the latter abject while the former is exalted.

The 'cartography'[19] I draw up, as such, concerns processes of formalisation and the valorisation of form. Jünger, as I will show, was a prolific writer obsessed with form: '*Form und Fassung*,' the latter translating as composure, countenance, cramp.[20] I am using Jünger to consider how the 'human' (the *Gestalt* of 'man') implicates itself, or is implicated, in the insectile as fascist fantasy assemblage, determined by a fetishization of total form and gestating out of an entomological imagination. To cast an eye or ear across to the other does not automatically, as it were, mean that the prevailing logic (of the Same) is broken or disrupted. Just because new figurations have the potential to emerge, that which is other/ed might well become part of, assimilate into, or be stamped out by, the ossature of a harder, frozen subject. *Homme-insecte*, emergent out of Jünger's writings, is enthralled by *Form und Glanz*, the shine [*Glanz*] of a fetishised form which corresponds to a valorisation, rather than the destruction, of form that defines Jünger's oeuvre.

In *Vampyroteuthis Infernalis*, a piece about the discovery of a hitherto unknown cephalopodic creature living in the depths of the South China Sea, Vilém Flusser (1920–1991) performs a mirror stage of sorts, intending to bring us to confront the vampyroteuthian trace in 'our' constitution. He writes that 'we are not attempting to vault out of the world but to relocate into another's' and conceives of his effort as a fable,[21] the genre with which Foucault starts *The Order of Things* (1966) to discuss, by way of a different system of thought, the limitations of our own.[22] Flusser's 'diabolical' deconstruction of the 'human'—diabolical derives from '*diaballein* = to cast across to the other side, into disarray'—is a metaphorical enterprise, so as to acknowledge and recognise something of the other in 'us'.[23] What is apparent in his article is the operation of a psychological motor, if you will, which more generally informs Flusser's proposal for a non-anthropocentric viewpoint and establishes the correspondence between biology and psychology (there already in Freud's work) through Reich, for whom the organism is a palimpsest, its 'stratified memory' like 'geological formations'.[24] Flusser's article, even though focussing on a transformation of a different order—the rationale for setting up this game of reflection is a critique of 'our' vertebrate existence from the 'perspective of the mollusc'[25]—prepares us for another correspondence ('human'←→insect) as well as for an engagement with the form of the 'human'. It offers a hinge I can use to pry open an analogous sociability, namely between the 'human' and the insect, which has a faint presence in Flusser's text. Notably, it emerges in his discussion of Reich, specifically in relation to a subject's character armour, which 'unexpectedly' leads to the impression that the insect body is 'the paragon of all organisms,' arrived at through Reich's decision to segment every organism into three parts: head, thorax and abdomen.[26] Each animal life form is, hence, an annulated being, cut into multiplicities. Etymologically, Derrida reminds us, insect comes from *inseco*, to dissect, to be 'divided into small strangulations by so many annuli'.[27] In *Character Analysis*, Reich writes that a subject's 'muscular armour is arranged in segments,' a structure found in 'a much more primitive form of living functioning,' like that of 'ringed worms,' but which nonetheless determines 'our' own nervous system.[28] He proceeds to analyse each armour segment in more detail in order to, ultimately, break it down—the point, after all, of undergoing psychoanalysis; the concept of the segment derives, in a first instance, from a study of the facial musculature:

> rigid forehead and eyelids, expressionless eyes or bulging eyeballs, mask-like expression, and immobility on both sides of the nose are the essential characteristics of this [particular] armour ring. The eyes peep out as from a rigid mask.[29]

Reich does not predominantly give bearings to the insect body as model for the 'human,' or any other animal organism, even if he mentions the caterpillar. The armour segments have a horizontal, circular structure through which 'orgonic' energy—pulsations that loosen the segmental armour rings—flows lengthwise; they recall a worm. As Reich notes: 'In the segmental arrangement of muscular armouring, we meet the worm in man [sic]'.[30] Accordingly, it is Flusser who perceives the insect in the armour rings, a body that remains unidentified but whose presence is nonetheless locatable there, even if Reich does not mention it (by name). It is indeed rather subtle, to be found, *viz*. Derrida, in the etymology of the word insect: it 'means *cut*, it names the cutting'.[31]

In a curious inversion, the endoskeleton is associated with rigidity and being locked into patterns of thinking and behaviour—the endoskeleton 'buttresses our bodies and attitudes toward life'—prompting Flusser to doubt the insect model's 'applicability to the human psyche'.[32] In other words, the insect is here coded as formless or in relation to its potential for formlessness, because it can divide and multiply, thus has little resonance with the cramped, endoskeletal 'human'. In this iteration, the insectile fits, much more closely, to *Vampyroteuthis*, which already 'belongs to a branch of life that derives from annelids,' for which 'segmentation is ingrained in its "collective unconscious"'.[33] And yet the insect remains allied to the 'human'; it is as if Flusser can't quite divest himself of the hold of its image, more specifically and finally its exoskeleton, returning in a particular 'form' or trajectory of orgonic energy. Concentrated in annelids, orgone nevertheless 'exploded' into two directions, including 'the direction of the armour, of militant rigidity and death,'[34] the realm, that is, of the insect conceptualised in terms of its shell, or what Rachel Murray calls 'ideas of enshellment'.[35]

The meaning of the insect body has changed considerably and unobtrusively: the fixed rigidity of the 'human' endoskeleton suddenly coincides with the insect's exoskeletal structure. Instead of being associated with the ring-shaped, the *insecta*, that which cuts and swarms, it now designates *thanatos*, the 'militant, [...] moribund and firm'.[36] On the one hand, then, the insect or insectile marks the formless, its etymological origins already giving a clue to its continuous dislocations, metamorphoses, the differences from itself. This, if you wish, is its 'molluscan' or vampyroteuthian function, defined by incessant movements of becoming resistant to form. On the other hand, it is indicative of the opposite, that is, total form: not segmentation but consolidation, armour, carapace, Hitlerian emergence, 'militant rigidity and death'. What we find in Flusser's work is the unresolved and unresolvable, as well as unremarked, tension between the two. Whereas he concentrates on tracing the former (vampyroteuthian) aspects of the insect/ile, which allows him to explore an eccentric, orgasmic experience of the world, I let the thanatoid perspective dictate and dominate the proceedings below. There is, as such, one rendition of the insectile that is not linked to becoming, to a dynamic, fluid, form-resistant subjectivity but is, on the contrary and evident in Jünger's work, linked to a purely static figure. I'm interested in finding out what happens if processes of crossing over into the dimensions of the other, rather than yielding an estrangement from form, enable its integration, its crystallisation, in the sense that these terms suggest density, coherence.

Disgust, repulsion, expulsion: these affects can be engineered, organised as products of historical or political forces or both, as Jünger in fact notes: the scarabaeus is not invariably a 'repelling creature' and functions differently in different contexts.[37] What Jünger does, across his oeuvre, is stabilise the insect as form and, more so, as a form of mythic speech (to refer to Roland Barthes), in which it is valorised, not negated. The valorisation itself might not appear as overly problematic, considering the disastrous global loss of insect populations, but as a form of attention it presupposes utility: 'useless forms' of life are still condemned to extinction. Above all, we need to interrogate how Jünger makes his insects mean, which is an endeavour that is not structured in terms of the other. The 'intolerable,' that which insists on difference, is thereby neutralised, and integrated into the system of meaning and desiring trajectory of the phallic I. This system revolves around the function of the *typus*, appearing as eternal form, exceptionally robust, able to withstand considerable pressure against that which arrives from outside, the namelessness [*das Namenlose*] as yet not captured by discursive language, categories of order:

> When we address a particular animal, such as an insect, as 'scarabaeus,' this is preceded by the encounter with a transient entity. We set and designate it as *typus*: the name [scarabaeus] now delimits a category in which we can easily accommodate all other specimen of this type, whether we encounter them in nature or not, even if they are only experiments in thought. Legions fit into it. *Typus* is the model by which we take measure.[38]

For Derrida, genre 'invaginates,' but such an imagination, characterised by mobility, impurity and exuberance, is inconceivable for Jünger: invagination presupposes that very namelessness or formlessness he is determined to control.[39] While the concept of accommodation might suggest some elasticity, the willingness to include 'legions,' the predominant focus clearly is delimitation, the policing of borders, the safekeeping of a 'purity' of identity and classification. *Typus* sets standards; it thrusts out of the obscene, into the *Glanz* of total form.

Typus

By total form, I direct attention to Jünger's 'ethos' of 'total mobilisation,' the 'marshalling' of all energies into a process of armouring that penetrates, and subsequently qualifies, the 'deepest marrow' [*eine Rüstung bis ins innere Mark*] of the subject.[40] The resulting impression of such a 'type' is a dense, in the sense of fully present, absolutely coherent figure, whose ego is exterior machine. The revolution of the *Gestalt* that Jünger imagines in *Der Arbeiter* [*The Worker*] (1932), as well as elsewhere, is rooted in entomology, itself pushing against that which is nameless and without order [*das Ungesonderte*]. In the space of his exhaustive writings and re-writings, the product of obsessively returning to earlier versions of his texts, he develops a distributed but no less consistent narrative about *Typus, Name, Gestalt*, to cite the title of one of his essays. *Typus* drives Jünger's conception of the worker, engendered as model species-subject and embodying a new will to power, total mobilisation and 'race-building' [*Wille zur Rassenbildung*].[41] *Typus*, an exalted form of entomological origin, is shaped by coleopteran resplendence.

In his beetle diaries, which function simultaneously as specimen records and scrapbooks [Figure 2.1]—Heike Gfrereis calls them 'convolutes'—Jünger lists the creatures he encounters and preserves.[42] On 11 March 1961 in Damascus, for example, he registers Tenebrionidae (darkling beetles), Carabidae (ground beetles), Scarabaeus (dung

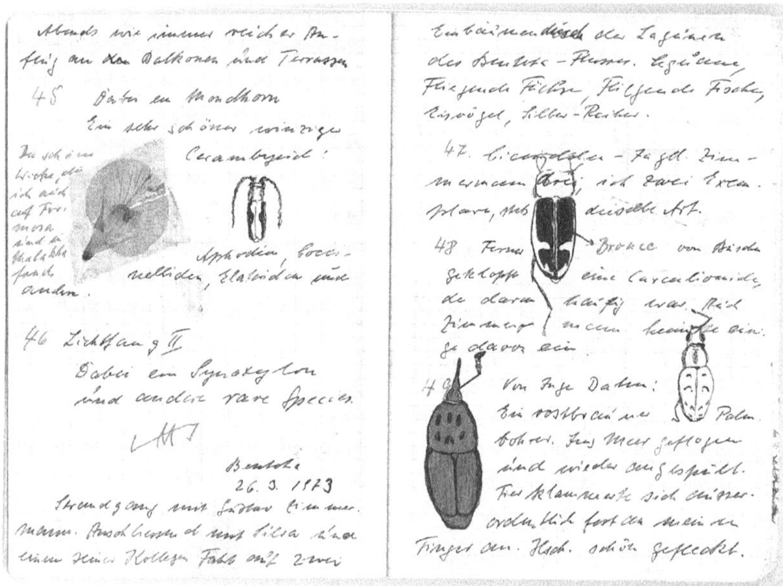

Figure 2.1 Ernst Jünger, page from *Coleopteren* IX © Deutsches Literaturarchiv Marbach.

beetles), Cerocoma (blister beetle), Ichneumonidae (parasitoid wasps), Zophobas (a species of darkling beetle), and Cymindis (a genus of ground beetle).[43] His *Coleopteren*, Jünger's term for these diaries, are teeming things, integrating various textual, visual and material elements or remnants: newspaper reports, postcards, stamps, dried leaves and flowers, the occasional mutilated specimen—squashed or missing limbs underneath pieces of tape; the formless *par excellence*—drawings of the coleoptera he observes on his travels. They are impressive documents, products of a taxonomy, an order of visibility as 'camera di morte,' never mind their apparent eccentricity, the ostensible disorder of their pages, which need to be looked at in the context of his concern with form and formatting.[44] The recording eye, killing machine, renders 'nature' as immobile, even more striking when considering that the objects, here, often fascinate because of their transformations, their ability to appear other than what they were before and will be after. In *The Order of Things*, Foucault notes that with 'Tournefort, [...] Linnaeus or Buffon' a 'new kind of visibility [was] being constituted in all its density,' by which he means a way of discovering and arranging forms, patterns, and so on, 'handed on down the centuries while preserving [a] strictly defined identity'.[45] The maintenance of form is, in many ways, the most important element: it is according to a *typus* that all variables of description are measured, captured in a 'system of names'.[46] As Foucault writes: 'the naturalist is the man [sic] concerned with the structure of the visible world and its denomination according to characters,' that which distinguishes one specimen from another and which must be confirmed in detail, at once providing a 'certain *designation* and a controlled *derivation*'.[47] It is a 'well-constructed language,' characterised through precision and precisely departing aspects subsequently assigned to an 'area of adjacencies'; natural history, in short, operates as 'possibility of a constant order'.[48]

The setting of desire for the collector, should she indeed deviate from the natural historian, similarly is order, precision, the idea of the proper: the proper name of a specimen, located in its designated place. Manfred Sommer, in his study on collecting, argues that a collector is interpellated as such through adopting the formal pattern—which bestows meaning onto each and every scene of collection—of a classificatory system.[49] This system, once taken up and rigorously practiced, sets rules to be heeded and obeyed. There is pleasure in such an activity—a dimension missing in Foucault's account—in the study of form and the diversions of form, though it also manifests a certain element of compulsion, of being under a spell, itself not necessarily distinct from the experience of pleasure. Sommer establishes language, the *fourmillement*—a swarming or multiplication of elements occurring around a particular term [*Begriff*]—as site of pleasure. If a term constitutes the standard according to which things are held in place, with respect to a *typus*, it further functions as an area where the adventures of reading and interpretation, of minute deflections, can emerge.[50] The system finally always turns out to be incomplete; the desire for, and impossibility of, totality determines the art of collecting. It is this paradox which provides an explanation as to the '*tremendum*,' which Sommer concurrently calls the '*fascinosum*,'[51] that accompanies the practice, absolutely marked, according to Lacanian algebra, by *objet petit a*, the 'punctiform, evanescent function' of desire.[52]

Fascination, as Lacan reminds us, is central to the organisation of the subject, which 'gains its unity in so far as it is fascinated'.[53] In other words, fascination is structured by fantasy (or desire), itself dependent on *objet a*. At once an event of surrender where the 'gaze gets taken in,'[54] is lost in an object, fascination concurrently is the scene where the phallus arises to ward off the threat of dissolution of form. In *The Space of Literature* (1955), Maurice Blanchot, concerned with the tantalising promise of being taken in, observes that fascination is blinding, absent presence 'fundamentally linked to [a] neutral, impersonal presence, to an indeterminate They, the immense, faceless Someone'.[55] The subject sinks into the light of this milieu, whose neutrality and impersonality emanate from a multiplicity or from some mysterious, immobilising entity: think, for example, of Dr Mabuse, the terrifying voice without body in Fritz Lang's *The Testament of Dr Mabuse* (1933). My stake, here, unlike Blanchot's, is to turn the gaze towards the defensive reaction, the stiffening phallus as emissary of form: the phallus announces form, the *Gestalt* of the 'proper' subject.

In the context of the collector-subject (though no different for anybody else, following Lacan), the field of desire is impelled by a 'well-ordered whole,' a 'plan, pattern, scene, matrix,'[56] so that the fascinating object, intervening into the system of the ego, is set up as a 'deceiving and idealised image' giving the subject its identity.[57] Sommer writes that the collector succumbs to the spell of a universe of order, upholding at its centre—what Derrida calls 'the point of presence'[58]—an evocation of form, of *typus*, as *arche*: immaculate specimen, it is origin and destination/destiny, the past and future form towards which all other forms refer.[59] The body in pieces assembles or collects itself as/into the image of the subject of mastery by handling—the word *Begriff* speaks to the importance of grasping [*greifen; Griff*]—objects to be arranged into a coherent structure, in turn organising the structuration of the subject. Fascination, consequently, is a structural relation, rather than a property attributed to specific elements (stamps; insects; records; coins; clippings of the German football player and Borussia Dortmund captain Marco Reus, etc.); that much is in fact already clear when considering that Sommer talks about a philosophy, not simply an object, of collecting. It is an art, then, all about processes of forming: linguistic, conceptual, concrete. An entire environment, including archives, filing systems, libraries, indexes, inventories, is devoted to the regime and maintenance of form and the enclosed play it enables.[60] The moment the subject-collector, discovering her unity in the very act of collecting, starts gathering objects, these must be held together in some form or other. The *Begriff*, more than anything else, is indicative of the scene of formalisation that is also taking place, in a room somewhere, in a glass case, in diaries, such as Jünger's (Figure 2.1), that break the linearity and two-dimensionality of their pages.[61]

According to Heike Gfrereis, Jünger disturbs the logic of writing in his diaries, at times approaching the aesthetic of a landfill; she further describes his manuscripts as palimpsests, in various states of aggregation [*Aggregatzustände*].[62] She considers dreamwork the procedure by which Jünger accumulates things on the page which constantly redefine the order or pattern of his manuscripts: layered, accrued, expanded upon in disappearing margins. These are themselves part of a collection occurring in the realm of writing, which, in the very process of assembling and 'revising' an order, always seems to threaten falling into chaos, to be overcome by, and turn into, kipple, Philip K. Dick's *novum* for objects degrading into trash. Gfrereis argues that, for Jünger, authorship does not mean obtaining mastery over his corpus of writing [*Werkherrschaft*][63] but that instead it records, even allows, encounters. This is a compelling perspective, one that perhaps resonates with a reader who, for her part, is first encountering Jünger's *Käfertagebücher*, keeping track of iridescent Tenebrionidae, chronicling their passage along paths crossed. Yet it is an approach that, though aware of how his writing stands in relation to the act of collecting—writing as collecting—fails to see how the diaries are committed to or strive for order [*um Ordnung bemüht*] and hence how they function as 'enormous order-machines.'[64] An entire apparatus of formalisation is at work, which, over the course of a century—Jünger lived to be over a hundred years old—also mobilises the myth of 'total authorship,'[65] a lifetime of productivity undertaken by a worker-subject, authorizing his life as pure work, the 'gigantic labour process'[66] marshalled in *Die Totale Mobilmachung* (1930).

It is difficult to see all this in terms of a *Werklandschaft*, a network of extensively correlated writings, that is not, at the same time, precisely a *Werkherrschaft*, an act of command, of being in command over worlds of labour, not least because Andreas Huyssen interprets the obsessive rewrites of Jünger's early work, especially, in terms of the concept of the armour. Palimpsests, at any rate and as Reich teaches us, can be indicative of armouring rather than of the world turning to kipple. Reflecting on the amended versions of Jünger's *In Stahlgewittern* (1920; by 1934, 14 editions had been released), Huyssen observes:

> All of Jünger's writing of the 1920s, including the revisions of *The Storm of Steel*, are marked by the attempt to forget that tiny, fragile human body, or rather to equip it with an impenetrable armour protecting it against the memory of the traumatic experience of the trenches. Forgetting as an obsessive rewrite project, with each additional layer of text another repression, another exorcism, another piece of the armour.[67]

The strategic function of writing, here, indeed concerns assembling: the assemblage of an armoured subject distinguished through his—because it is, without a doubt, a phallic subjectivity—'metallic coldness'.[68] There is, however, something else to comment on, considering the figure galvanised throughout Jünger's oeuvre, erecting itself in opposition to that which is vulnerable, formless, has the potential to disintegrate. The latter, of course, is the precondition for a new form to arise, an event Jünger repeatedly describes in *Der Arbeiter*, where he often also uses a vocabulary deriving from hard labour in conjunction with the deployment of an entomological gaze. There is a constant switching between these two formalisations, one 'natural' and the other the result of *technē*; in either case, form is the privileged condition. A *typus* arises out of a *Gestaltwandel*, a metamorphosis which entails a melting out of those aspects that prevent the purity of a metal, a transformation that is naturalised later on in *Der Arbeiter* and linked to the image of 'life consum[ing] itself'.[69] The transformative process is likened to a caterpillar's evolution into the imago: a caterpillar, after a phase of rapacious consumption elongating and swelling its body, repeatedly moults, sheds its skin, before digesting itself inside the chrysalis it forms. The metamorphosis that occurs passes through various stages of form and formlessness, the caterpillar's tissues dissolving to leave only 'imaginal discs,' cells that contain the data for the imago, the future form.[70] The *Gestalt* of the worker—emergent after a period of loss, the melting or moulting of the 'impure,' which in turn resembles the sculpting of a statue, losing the mass of stone out of which it is cut[71]—clearly derives from the caterpillar's biological journey, even if continuously slipping into *technē*. What is discarded, in this worker-imago, is the formless, that which resists the character impression of the new *typus*, which distinguishes itself, above all else, in the face:

> The face staring back at its observer from under the steel helmet or the pilot's cap has [...] changed. The range of its expressions has diminished in its multiplicity and thus in its individuality, as can be seen in a gathering or in group photographs, while it has gained in the precision and specificity of its singular features. It has become more metallic, galvanized on its surface, so to speak; the bone structure is prominent, the features sparse and taut. The gaze is quiet and fixed, trained in the observation of objects apprehended under conditions of high speed. This is the face of a race which has begun to develop in response to the particular challenges of a new landscape and which the individual represents not as a person or as individual, but as *typus*.[72]

The face mask, recalling Reich, announces a much more extensive *Verwandlung* into cold figure, to refer to Helmuth Lethen's work, 'operating under the spell of total mobilisation'.[73] These appearances—mask, *typus*, *Gestalt*—are aligned despite the divergences that Jünger elaborates on elsewhere, in that they overwhelmingly resonate with what Elias Canetti calls the 'stereometric figure'. In *Crowds and Power* (1960), Canetti describes this figure as an angular subject, a 'soldier like a prisoner who has adapted himself to the walls enclosing him' and are 'affect[ing] his shape'.[74] It is, similarly, a figure evoked throughout Klaus Theweleit's *Male Fantasies*, Vol. II (1978), where it is synonymous with a singularly finished form, going under different names: *Stahlgestalt* (figure of steel), a utopian totality-component because at once whole in, and fully present to, itself and frictionless part of the troop; 'child of the drill-machine'; fantasy-man whose 'physique has been machinized, his psyche eliminated' or 'displaced into his body armour,' realising a 'machine-like periphery'.[75] The ego, as Theweleit demonstrates, is a 'muscle-physique' stabilised against collapse through work or, rather, as Werner Hamacher argues, through a particular and pervasive conception of work intent on exterminating 'the nonhomogeneous, the nonassimilable, [...] the formless'.[76] Hamacher shows that as long as the 'endogamous fascism' of such an interpretation of work keeps evading analysis, it behaves as one of the many ideological, social and political continuances of the Nazi regime, a system that 'defines itself as the rejection of what is foreign to work and the foreignness of work "itself" through murder.'[77] What he means is this: work is form and formalization; it corresponds to the imposition of form and the elimination of the formless, that which is coded as other, can't be gathered into form, might be form-resistant, has the propensity to deform. Work, in this sense, is anti 'anti-form,' never mind that the formless makes form possible. That disavowal, the ability to grant the formless the right to exist, irrespective of whether it can be constituted into form, structures fascist ideology while persisting in organising the current capitalist mode of production or, more precisely, racial capitalism. Seb Franklin has analysed the 'discursive maintenance of states of form and formlessness,' which sustains an economic system, a racist-capitalist imaginary, predicated on the valorisation of form.[78] This logic of form—a political and epistemological condition that remains ongoing—crushes the formless. Work pressed into the service of this logic is synonymous with murder, the absolute dispossession of rights or recognition of 'forms' of life that do not correspond to the 'good form,' the *Gestalt* of 'man' monumentalised against an abjected formlessness.

Gestalt, typus, stereometric figure = the 'good form,' distinguished through pure, definite lines; a phallic, vertical I; an image of wholeness, a subject without remainder. Even though this figure requires constant maintenance—hence the enormous amount of fascist writing, behaving as mechanism assuring stability[79]—it is a static subjectivity which, through endless processes of revision, attains its ultimate, 'proper' form. The economy of work, and at work, here, a process of assembling an armoured form, yields a *Gestalt* whose articulation despite or because of its constant reiterations is metallic, machinic, mythic, immobile. In this context, Jünger's amendments of his own texts are indicative of such a process of forming, which he describes as a sloughing off, skin discarded after undergoing ecdysis.[80] What emerges or hatches is a fantasy whose name might be legion—*Stahlgestalt*, stereometric figure, etc.—but whose form is that of the insect-body as *typus*, prototype or paragon of the armoured subject. Metamorphosis, as such, matters only in terms of its final stage, the codification of Being as insect/ile: exoskeletal ego, hardened into a carapace; no secret interior; impassive face; 'proper' or total form that has reached completion and gained its imago. Jünger seizes the (idealised) subject as *homme-insecte*, whose metallic sheen is at both the origin and the end result of a development that proceeds to fully incorporate what Deleuze and Guattari call a 'molar' entity, the single, identifiable, unified, authoritative subject.[81] In this respect, there is indeed a dreamwork expounded across Jünger's writings, even if it effectively banishes Freud (the subject, after all, is pure ego). It is dreamwork as wish-fulfilment, dependent on the functioning of writing as hatching, excoriation or, following Connor, entomo-oneirism: a 'dream of the insect, according to which a confining outer skin must be split in order to give room for the new body to emerge'.[82] It is not simply, then, that the notion of *typus* derives from entomology but also that the very act of writing is entomophilic; it dreams of hatching an armoured self, all shining skin ➔ a new resplendent mode of being.

Fetish

The fetish is a carapace: it is a fixed thing—it endures; it has an originating event—and a thing which fixes, has ordering power.[83] In the first section to 'The Problem of the Fetish,' William Pietz cites Michel Leiris, the French writer and ethnologist, who argues that fetishism is narcissistic love or infatuation projected outwards, where it hardens, takes on form. There, outside, the 'solid carapace [...] imprisons [this love] within the limits of a precise thing and situates it, like a piece of furniture which we can use in that strange, vast room called space.'[84] A territorialised object, a piece of furniture in a room, the fetish is also sign of a displaced lack whose 'presence' elsewhere is surrogate or substitute phallus. In the final part of this annulated chapter, I am tracing a trajectory of an inside (a narcissistic love, a lack) projected outside and crystallising as coleoptera, standing as 'memorial' to a 'perfectly finished' subject without 'hint of lack or loss'.[85] To do so involves being alert to the stag beetle as fetish object mobilised by a discourse whose order-word is fascism, a logic assembled or remembered in the experimental documentary film *Vom Hirschkäfer zum Hakenkreuz*, concerned with the use and manipulation of images. More specifically, it traces the cooperation between avant-garde filmmakers, often working on nature documentaries (*Kulturfilme*), a genre that despite its 'conservative' focus played with form, and the Nazi regime.

In *Subtile Jagden*, the autobiographical account of his insect collecting, Jünger recounts the start of his obsession, bequeathed by his father. It is a heritage, passed down from father to son, that recurs: in *Vom Hirschkäfer zum Hakenkreuz*, a *Historionaut*—history machine, it is also an associative machine, a montage and dream machine, a search engine—gathers resonances between (grand)fathers, stag beetles and avant-garde cinema before leaping to cyberbugs and Dark Wave music of the 1990s. The film's argument is implicit, its quasi-somatic method transposed and developed, using Jünger as a decoding key. Its driving force or the machine's command, which dials itself into networks, is the production of the seductive image, even if the narrator-machine begins by giving voice to a silent 1921 film, *Der Hirschkäfer*, whose first intertitle sets the scene: a morning walk in an oak forest. Two men, sucking on pipes, are on the lookout for stag beetles, delivered in macro-optic close-ups, a technology which Ulrich K. T. Schulz, the director, was the first to use. The beetles are discovered in bleeding vaginal folds of oak trees; the narrative, both visual and through intertitles, calls attention to the enormous mandibles (themselves phallic substitutes) of the male stag beetle. A patriarchal, pseudo-Darwinian imagination is at work: female stag beetles can 'justifiably be called the weaker sex'; the beetles' short life span, the summer months, frequently ends in battle: '*Sieg—und—Tod*,' victory and death.[86]

The *Historionaut* explains a tradition of beetle-collecting trips to oak forests with 'granddaddy,' who also made educational films called *Kulturfilme*, initiated in Germany after World War I. Between 1918 and 1945, 20,000 *Kulturfilme* were produced; they were screened in cinemas as previews to main features, and their focus, especially in Nazi Germany, was the stag beetle in its struggle 'in and against nature,' an 'inexhaustible, eternal subject' in fascist thought.[87] *Der Hirschkäfer* is brought into connection with *Das Erbe*, a 1935 Nazi propaganda film using fighting stag beetles recruited to the discourse of 'racial policy' set forth by a 'professor' facing an idealised Nazi *typus*, a beautiful, young, blonde woman: 'in battle, everything weak is eliminated. [...] Otherwise, all life would collapse due to its own infirmity.' *Vom Hirschkäfer zum Hakenkreuz* follows the stag beetle across its discursive iterations, from its occurrences in *Kulturfilme* to its functioning in the Nazi apparatus and from there—the unspoken link is Jünger—to Dark Wave. Notably, coleoptera are mentioned in relation to a 1996 Leni Riefenstahl tribute album curated by Josef Klumb, a far-right German musician whose one-time band project, Von Thronstahl, incorporates the black sun—a wheel consisting of radial runes citing the swastika and the SS symbol—in its logo. The beetle also appears, more obviously, on a track titled 'Käferlied' (Beetle Song), a tribute to Jünger[88] by the far-right Austrian band Allerseelen. An unremarkable, monotonous piece of about six minutes, the track's spoken words only occasionally emerge out of a sonic force field that, in its opening stages, stirs an apparent multiplicity. There is, however, only one signifier here: a single entity, the 'very last' beetle, a sacred object apparently lost, which has 'protected' and 'shielded him'—that other, commanding signifier—and must be recovered in the underworld.[89] Underworld-*Reich* and 'home,' to which he returns, coincide in this curious last-man narrative, in which beetle and man exist in correspondence: the passage into the underworld turns him into holy object preserved by 'dark beetles,' nameless creatures taking him up into their collection. The accompanying video, by contrast, shows not a man but a woman, an irradiated dancing figure, whose veins and luminous triangle indicative of her reproductive organs are more distinctive than the contours of her body: fertile *Blut und Boden* of an interior corporeal landscape.[90] At times, two of these ghostly shapes arise to merge back into each other, roughly arranging into an x-rayed close-up of the over-sized mandibles of a male stag beetle, a coming together obscurely pointing back to the set of coordinates utilised by the Nazis: a nexus of 'sacred' and tragic being; 'pure' blood in unobstructed flow suggested by the insistent dance sequence; and instantaneously visible, fully present, phallic power.

A mythical (as well as mystical or occult) system is in operation, consisting of elements assembling the stag beetle into relations with the impression of arresting size, oak tree/forest, father figures, heritage, fertility, soil, blood, battle and victory, eternalising a message about nationhood, patriarchy, naturalness, strength. The order of signification above invokes the armour, the sheen of the insect's exoskeleton, further referenced, or incubated, in the musical genre, which Stéphane François describes as 'euro-pagan,' an identitarian, *völkisch* scene marked by esotericism and, in certain cases, 'SS-occultism'.[91] With minor key, low pitch and droning compositions, contiguous to things buried, close to, or below, earth, Allerseelen—whose name translates as All Souls' Day—are a neo-folk, euro-pagan, post-industrial and martial industrial band; this musical genre, according to Anton Shekhovtsov, is often 'apoliteic,' an orientation closely associated with the European New Right (ENR).[92] The apoliteic is characterised by distance from the 'modern' world and its 'values,' thereby veiling its alignment with fascism, and hedges around Jünger's essay and notion of the *Waldgang* (1951), a walk and retreat into the forest. This 'interregnum' of sorts—awaiting the palingenesis of a highly mythologised Europe—is fascist, because a form of revolutionary ultra-nationalism dreaming of a new, spectacular temporality: a 'secret Europe' to be generated in the forest functioning as symbol of 'an enduring organic rootedness'.[93] It is not only the apoliteic and/or the 'metapolitically fascist'—that is, not engaged in violently implementing fascism as political system—that is discovered, however, as intimately constitutive of this particular musical genre, because its laws are further, and as intimated above, cognate with the insectile, not least in terms of its semiotics. Industrial music renders static sounds and timbres—buzzing, hissing, the *zoē* of swarm 'noise'—that summon the insectile, the formless iteration appearing in the name/figure of that new, still occulted, Europe carried on the backs, also, of stag beetles as (crypto)fascist vehicles.

As a mythologised object or message—it is difficult to talk of entities when it comes to sounds and swarms, 'bodies without surface'[94]—the stag beetle is a motif incorporating the curious 'interregnum' and the discourse of the organic + the technological (as well as form and formlessness) united in the same (cyborgic) organism/aggregate. Associated with earth, oak, burrowing, it is equally bound up with *technē* and armouring, not only with respect to the music analysed here but also regarding the *Kulturfilm*, indicative of 'style, taste, form and accomplishment [*Fertigkeit*],'[95] itself a set of terms whose body of associations encompasses completion, immobility, perfection. The mode of presence of the beetle, condensed through the speech of myth in both genres—martial industrial, euro-pagan music and *Kulturfilm*—hence includes and gestures to the machinic, implied by sounds and/or glistening in close-up shots: exoskeletal 'dark plates' that, 'wonderfully shaped' and geometrically precise, to cite William Gass's 1968 short story 'The Order of Insects,'[96] become abstraction, arrestation, harmonious form. In Gass's story, a suburban housewife becomes enthralled by dead bugs, whose features hold in death as they did in life: they are the epitome of form preserved eternally. This order—gracious and static—is moulded across the material discussed above, in which the stag beetle is abstracted from its 'natural' environment and deployed as 'nebula,' as it were, a 'condensation' of elements that intends to blind or rather that shines, is designed to seduce.[97] In other words, it is produced as object of fascination, inviting and commanding desire. On a fundamental level, the desire that the stag beetle, as image, myth, and invocation, compels is the desire for the phallus, the fantasy of power, of being in possession of *objet petit a*. In *What Do Pictures Want* (2005), W.J.T. Mitchell argues that

> the process of pictoral seduction [Michael Fried, an art historian and critic] admires is successful precisely in proportion to its indirectness, its seeming indifference to the beholder, its antithetical 'absorption' in its own internal drama. The very special sort of pictures that enthral him get what they want by seeming not to want anything, by pretending that they have everything they need.[98]

The meaning that emerges from the image of the stag beetle—immune to our gaze, whose impotence is here exposed, often resulting in other displays of violence to crush the other's utter indifference, its opacity—is one of completion, of being absolutely untouched by the trauma of castration. At the same time, it hails and maintains related desires, having to do with loss (the loss of a 'true' Europe, etc.), beyond this most arresting one, mesmerizing the beholder who, lacking, stands transfixed by a figure of plenitude.

In 1998, as part of a travelling exhibition, a glass case in the Natur-Museum in Luzern, Switzerland, displayed a cross-section of a beetle next to the arm guard of an armour; both are distinguished through their splendour, the *Glanz* of their armoured form.[99] The insect (not limited to the stag beetle) is brought into the light as armoured thing, as site of desire organised around its shell as fetish object. In his paper on fetishism, Freud notes that the fetish is not initially perceived by the patient as a symptom linked to suffering and severe injury and that, through the case which opens Freud's study, it is linked to light by way of the German word *Glanz*, wrongly lodged in his patient's memory. *Glanz* had to be understood in English, as glance, writes Freud, because his patient had moved from England to Germany: the nose, here, operating as fetish object resplendent with a 'luminous shine'.[100] The fetish, Freud continues, is coupled to trauma, the 'horror of castration,'[101] despite its resplendence, and is at once disavowal and affirmation of loss; it is designed to preserve from extinction, to protect the ego against the terror of lack. To be brought into the light, as such, frequently means to mobilise an object as fascinating and, concurrently, as fetish; as Connor shows, the 'desire for fascination is a desire for arrest, but of a certain enlivening kind, in which the subject of fascination is at once enthralled and aroused.'[102] This particular phenomenon, of being 'arrested into arousal,'[103] asserts the presence of the erect phallus (as well as the function of the fetish), intending to defend against threat. While Freud dismisses 'scotomization'—by which a perception is totally wiped out, as if an impression fell on a blind spot in the retina—the play of light is something that remains, is retained in the fascinating relation between subject and object.

Accordingly, in *Subtile Jagden*, Jünger's fascination—it is indeed articulated as such—with insects far exceeds his father's predilection, noting that the 'small objects gradually acquired a magical shine [*Glanz*]'.[104] In a lengthy passage, he explains the different affects emanating from, on the one hand, butterflies and, on the other, coleoptera, commenting on a 'voluptuous' sense of pleasure and delight generated by a butterfly's beating wings. (Wolf Man comes to mind, seized not by pleasure but by an irrational fear at the sight of a butterfly, whose beating wings trigger the patient with impressions of a woman's spread legs, thereby perhaps catching sight of, or suspecting, her 'lack'.) Jünger remembers first beholding a Morphos, a type of butterfly, whose wings, when shut, shone like gold brocade; when open, they resembled a silver-coated mirror with a sky-blue base. The apparition increases its charm—Jünger emphasises the stillness of the scene, taking place under a burning sun—like the glance or gaze of an eye that, with each opening and shutting, is put more conclusively to sleep [*der Bann wurde stärker und stärker, wie der Blick eines Auges, der vom Lidschlag immer mächtiger, immer zwingender einschläfert*].[105] Jünger sets up a correspondence between the blinks of an eye and the movement of the creature's wings, both of which beat, eye mirroring wing and vice versa. The beating is shared, fully attributed to neither the one (subject) nor the other (object) but discovered in a radiant space in-between. Beauty, he proceeds, robs us of the proper, of what is 'proper' to 'us,' an event explicitly formulated as threat and pleasure at the same time. Fascination, rather than a 'transitive phenomenon,'[106] producing an agent and a recipient, a vessel, as it were, is that which shifts the I into the spell and sphere of the other but without ascribing all the power to an entrancing object. The subject, after all, is a desiring subject, whose gaze or glance projects *Glanz*.

Jünger opposes the materiality of the coleoptera to the gracefulness of the butterfly in *Subtile Jagden*:

> They are of harder matter [*stoffharter*], harder and, as jewels of the earth, aligned more obviously with fruit than petals, conches and

> crystals instead of being kindred to birds. They do not reveal their beauty all at once, which means that their admirers [*Liebhaber*] tend to be more constant [*beständig*] than those who value butterflies.[107]

A vocabulary of hardness prevails: it insists through repetitions, the evocation of hard matter (jewels, conches, crystals) attracting harder lovers [*Liebhaber*], whose *Beständigkeit* codes not only constancy but also the stiffened phallus, glimpsed in the word *ständig* and therefore *Ständer*—the erect penis. Coleoptera seem, at times, to be inanimate, less related to living beings like birds and more akin to shells, hardened, crystalline objects, things of the earth. Jünger keeps being drawn to their 'metallic shimmers' [*Metallglanz*], the 'lustre of [their] armour' [*Glanz der Rüstung*], commenting on the sheen of beautiful forms, subsequently arranged in his notations centring objects, giving them *Beständigkeit.*[108] Despite their brilliance, their emergence out of the *Ungesonderte*, these objects are and remain, or so the subject needs to ascertain, measurable, things to be captured. In *Typus, Name, Gestalt* (1963), Jünger elaborates on the formation and formatting of the unformed. To give a name is to insert an object, 'falling into a word' out of the nameless [*das Namenlose*], into a framework: through this act of restriction [*Begrenzung*], we delimit and recover territory from the sphere of influence of an amorphous namelessness.[109] The nameless is that which awaits being made into form yet which lurks beyond each image—the relation between form and formlessness is a dynamic one—and is to be counteracted by the phallus.

Projections of hardness occur, as such, in two places: at the level of the fascinating object—arising from 'outside,' the *Ungesonderte* out of which they grow, are hoisted[110]—and the entranced subject, who, at the sight of said object, turns stiff with terror. In 'Medusa's Head' (1922), Freud notes that the stiffening 'offers consolation,' in that it reasserts the viewer of the presence of his penis: look, I am still here, I am standing up. 'I am not afraid of you. I defy you. I have a penis.'[111] Medusa, of course, terrifies because she apparently, according to a devastating patriarchal logic, lacks. Coleoptera, as figures of plenitude and phallisized objects, on the other hand, return that lack where it belongs, testifying to the fact that the mesmerized subject does not possess the phallus, that the presence that manifests itself in response to their *Gestalt* is only ever 'dubious,' as Connor puts it.[112] They seduce, then, because they are in possession of the 'eternally lacking object'[113] from the perspective of the mutilated subject, always without *objet a*. But, rather than losing itself in this milieu, the scene or reminder of its own castration, the subject, bracing itself, supplements its iteration through the other, which it integrates as part of its armour. *Homme-insecte*, as it emerges in this space, is *homme fasciné* and *homme fasciste*, consumed by the obsession to re-member its ego-armour, the resplendent, mythic figure of the Ideal-I.

Notes

1 Joyce Cheng, 'Mask, Mimicry, Metamorphosis: Roger Caillois, Walter Benjamin and Surrealism in the 1930s,' *Modernism/Modernity*, Vol. 16, No. 1 (January 2009), 62.
2 Cheng, 'Mask, Mimicry, Metamorphosis,' 64.
3 Cristopher Hollingsworth, *Poetics of the Hive: Insect Metaphor in Literature* (Iowa City: University of Iowa Press, 2001), 195 & 200.
4 Jacques Lacan, 'The Mirror Stage as Formative of the Function of the I as Revealed in Psychoanalytic Experience,' in *Écrits: A Selection*, trans. Alan Sheridan (New York & London: Routledge, 2001), 3.
5 The reference is to Andreas Huyssen's article, 'Fortifying the Heart—Totally: Ernst Jünger's Armored Texts,' *New German Critique*, No. 59 (1993) Special Issue on Ernst Jünger, 3–23.
6 Gilles Deleuze & Félix Guattari, *A Thousand Plateaus: Capitalism and Schizophrenia*, trans. Brian Massumi (London & New York: Continuum, 2004), 8.
7 Virginia Woolf, *Three Guineas* (New York: Harcourt, Brace and Co., 1938), 80.
8 The German *Typus* translates as type in English. I keep it in German—which Bogdan Costea and Laurence Paul Hemming also do in their latest translation of *The Worker*—because I'd like to draw attention to it as site of pressure, as well as retain its speculative dimensions. Costea and Hemming further include the term in their glossary, explaining that *Typus* is as unusual in German than in English, and that it frequently recurs in Nietzsche's work, whose influence on Jünger's *The Worker* they wanted to highlight. See Ernst Jünger, *The Worker*, trans. Bogdan Costea and Laurence Paul Hemming (Evanston: Northwestern University Press, 2017), xiv, xv & xxvii.
9 Dan Gorenstein, 'Entomologische Horizontverschmelzung: Ernst Jüngers Hermeneutik der Käfer,' in *Inhalt: Perspektiven einer Kategorie Non Grata im Philosophischen Diskurs*, ed. Daniel Alder, Markus Christen, Jeannine Hauser & Christoph Steier (Würzburg: Königshausen & Neumann, 2015), 177. [All translations mine.]
10 Jünger, *The Worker*, 70.
11 Steven Connor, *The Book of Skin* (London: Reaktion, 2004), 178.
12 Since this paper is explicitly concerned with an armoured subject coded as masculine, I am sticking with the pronoun 'he' and 'his'.
13 Wilhelm Reich, *Character Analysis*, trans. Vincent R. Carfagno (New York: Farrar, Straus & Giroux, 1990), 115.
14 Vilém Flusser, *Vampyroteuthis Infernalis*, trans. Valentine A. Pakis (Minneapolis & London: University of Minnesota Press, 2012), 28.
15 William Pietz, 'The Problem of the Fetish I,' *Anthropology and Aesthetics*, No. 9, Spring 1985, 7.
16 Jessica Burstein, 'Waspish Segments: Lewis, Prosthesis, Fascism,' *Modernism/Modernity*, Vol. 4, No. 2 (1997), 156.
17 Burstein, 'Waspish Segments,' 157.
18 Lacan, 'Mirror Stage,' 3.
19 Deleuze & Guattari, *A Thousand Plateaus*, 13.
20 Ernst Jünger, 'Typus, Name, Gestalt,' in *Sämtliche Werke* 13 (Stuttgart, Germany: Klett-Cotta, 1981), 94. [All translations mine.]
21 Flusser, *Vampyroteuthis Infernalis*, 38.
22 Michel Foucault, *The Order of Things: An Archeology of the Human Sciences* (London & New York: Routledge, 2002), xvi.
23 Flusser, *Vampyroteuthis Infernalis*, 26.

24 Flusser, *Vampyroteuthis Infernalis*, 27.
25 Flusser, *Vampyroteuthis Infernalis*, 10.
26 Flusser, *Vampyroteuthis Infernalis*, 28.
27 Jacques Derrida, 'Fourmis,' in *Hélène Cixous Rootprints: Memories of Life & Writing*, ed. Hélène Cixous & Mireille Calle-Gruber, trans. Eric Prenowitz (London & New York: Routledge, 1997), 120.
28 Reich, *Character Analysis*, 756.
29 Reich, *Character Analysis*, 758.
30 Reich, *Character Analysis*, 762.
31 Derrida, 'Fourmis,' 121.
32 Flusser, *Vampyroteuthis Infernalis*, 29 & 11.
33 Flusser, *Vampyroteuthis Infernalis*, 29.
34 Flusser, *Vampyroteuthis Infernalis*, 29.
35 Rachel Murray, *The Modernist Exoskeleton: Insects, War, Literary Form* (Edinburgh: Edinburgh University Press, 2020), 13.
36 Flusser, *Vampyroteuthis Infernalis*, 28.
37 Ernst Jünger, 'Prometheus—Von Der Physik belehrt? Zu Wolfgang Wiesers Besprechung von *Typus, Name, Gestalt*,' in *Sämtliche Werke* 13, 174. [My translation.]
38 Jünger, *Typus, Name, Gestalt*, 99.
39 Jacques Derrida, 'The Law of Genre,' trans. Avital Ronell, *Critical Inquiry*, Vol. 7, No. 1 On Narrative (Autumn 1980), 59 & 57.
40 Ernst Jünger, 'Total Mobilization,' in *The Heidegger Controversy: A Critical Reader*, ed. Richard Wolin (Cambridge, MA & London: MIT Press, 1993), 126. For the original German, see 'Die Totale Mobilmachung,' in *Sämtliche Werke* 7, Betrachtungen Zur Zeit (Stuttgart: Klett-Cotta, 1980), 126.
41 Ernst Jünger, 'Der Arbeiter,' in *Sämtliche Werke* 8, Der Arbeiter (Stuttgart: Klett-Cotta, 1981), 110. The English translation uses 'will to build a race' in this case. See Jünger, *The Worker*, 69.
42 Heike Gfrereis, 'Schreiben als Träumen: Die Einklebungen in Ernst Jüngers Tagebüchern und Manuskripten,' in *Freiheit: Jünger Studien* 7, ed. Georg Knapp (Tübingen: Attempto Verlag Tübingen, 2015), 38. [All translations mine].
43 Ernst Jünger, *Coleopteren I*, 09.03.1961–12.12.1961, *Deutsches Literaturarchiv Marbach*, Marbach, Germany.
44 Ernst Jünger, 'Subtile Jagden,' in *Sämtliche Werke* 10 (Stuttgart: Klett-Cotta, 1978), 92.
45 Foucault, *The Order of Things*, 144 & 145.
46 Foucault, *The Order of Things*, 172.
47 Foucault, *The Order of Things*, 151.
48 Foucault, *The Order of Things*, 151 & 172.
49 Manfred Sommer, *Sammeln: Ein Philosophischer Versuch* (Frankfurt am Main: Suhrkamp, 1999), 24. [All translations mine.]
50 Sommer, *Sammeln*, 78. This discussion is further indebted to Derrida, *Fourmis*, 119–120.
51 Sommer, *Sammeln*, 80.
52 Jacques Lacan, *Four Fundamental Concepts of Psychoanalysis*, trans. Alan Sheridan (New York & London: W.W. Norton & Co.,1998), 77.
53 Jacques Lacan, *The Seminar of Jacques Lacan: The Ego in Freud's Theory and in the Technique of Psychoanalysis, Book II*, ed. Jacques-Alain Miller, trans. Sylvana Tomaselli (Cambridge, New York, New Rochelle, Melbourne & Sydney: Cambridge University Press, 1988), 50.
54 Maurice Blanchot, *The Space of Literature*, trans. Ann Smock (Lincoln & London: Nebraska University Press, 1989), 31.

55 Blanchot, *The Space of Literature*, 32.
56 Sommer, *Sammeln*, 82.
57 Lacan, *The Ego in Freud's Theory*, 54.
58 Jacques Derrida, *Writing and Difference*, trans. Alan Bass (New York & London: Routledge, 2001), 352.
59 Sommer, *Sammeln*, 89.
60 Foucault, *The Order of Things*, 143.
61 Gfrereis, 'Schreiben als Träumen,' 39.
62 Gfrereis, 'Schreiben als Träumen,' 40 & 44.
63 Gfrereis, 'Schreiben als Träumen,' 44.
64 Stefan Schlak, 'Ernst Jünger im Archive gelesen: Sechs Stereoskopien,' in *Ernst Jünger: Arbeiter am Abgrund*, ed. Karlheinz Bohrer, Heike Gfrereis, Helmuth Lethen & Stephan Schlak (Marbach am Neckar: Deutsche Schillergesellschaft, 2010), 25 & 26. [All translations mine.]
65 Schlak, 'Ernst Jünger im Archive gelesen,' 45.
66 Jünger, 'Total Mobilisation,' 126.
67 Huyssen, 'Fortifying the Heart,' 12.
68 Jünger, *The Worker*, 69.
69 Jünger, *The Worker*, 8 & 75.
70 Ferria Jabr, 'How Does a Caterpillar turn into a Butterfly,' *Scientific American*, 10 August 2012. Accessible via: https://www.scientificamerican.com/article/caterpillar-butterfly-metamorphosis-explainer/ [Accessed 10 July 2020].
71 Jünger, *The Worker*, 75. The English version translates 'abschleifen' as 'eroding,' losing the aspect of labour, which recurs in the next paragraph and page, where Jünger mentions the 'sculpted' body [*gezüchteter Körper*]. For the German version, see *Der Arbeiter*, 125.
72 Jünger, *The Worker*, 69. [My emphasis].
73 Helmut Lethen, *Cool Conduct: The Culture of Distance in Weimar Germany*, trans. Don Reneau (Berkeley, LA & London: University of California Press, 2002), 104.
74 Elias Canetti, *Crowds and Power*, trans. Carol Stewart (New York: Continuum, 1973), 312.
75 Klaus Theweleit, *Male Fantasies Vol II, Male Bodies: Psychoanalysing the White Terror*, trans. Erica Carter & Chris Turner (Minneapolis: University of Minnesota Press, 1989), 143, 159, 160 & 161.
76 Werner Hamacher, 'Working through Working,' trans. Matthew T. Hartman, *Modernism/Modernity*, Vol. 3, No. 1 (January 1996), 25.
77 Hamacher, 'Working through Working,' 29.
78 Seb Franklin, 'The Context of Forms,' *World Picture*, Vol. 11, Summer 2016, 2.
79 Theweleit, *Male Fantasies Vol II*, 228.
80 Jünger cited in Gfrereis, with Ellen Strittmatter & Sonja Lehmann, 'Die Ausstellung,' in *Ernst Jünger: Arbeiter am Abgrund*, ed. Bohrer, Gfrereis, Lethen & Schlak, 201. [My translation.]
81 Deleuze & Guattari, *A Thousand Plateaus*, 30.
82 Steven Connor, *The Book of Skin* (London: Reaktion, 2004), 178.
83 Pietz, 'The Problem of the Fetish I,' 7.
84 Pietz, 'The Problem of the Fetish I,' 11.
85 Hal Foster, 'The Art of Fetishism,' in *Fetishism as Cultural Discourse*, ed. Emily Apter & William Pietz (Ithaca & London: Cornell University Press, 1993), 261 & 262.
86 *Der Hirschkäfer*, dir. Ulrich K. T. Schulz (Berlin: Universum-Film AG (UFA), 1921), VHS.

87 *Vom Hirschkäfer zum Hakenkreuz*, dir. Madeleine Dewald & Oliver Lammert (Hamburg: Dock 43, 2002), DVD.
88 Rose City Anti-Fascists, 'Austrian Far-Right Musical Project on West Coast Tour,' *Who Makes the Nazis?*, 14 December 2010. http://www.whomakesthenazis.com/2010/12/rose-city-anti-fascists-austrian.html [Accessed 27 November 2019].
89 Allerseelen, 'Käfer Lied,' Track Number 7 on *Stirb und Werde* (Aorta, 1999). Lyrics accessed via Allerseelen Facebook page: https://www.facebook.com/allerseelen.allerseelen/posts/10153644563784739/ [Accessed 5 November 2019].
90 Allerseelen, 'Käfer Lied,' YouTube, 8 February 2014: https://www.youtube.com/watch?v=Cg2pLF49RTc [Accessed 5 November 2019].
91 Stéphane François, 'The Euro-Pagan Scene: Between Paganism and Radical Right,' trans. Ariel Godwin, *Journey for the Study of Radicalism*, Vol. 1, No. 2 (2007), 40.
92 Anton Shekhovtsov, 'Apoliteic Music: Neo-Folk, Martial Industrial and "Metapolitical Fascism,"' *Patterns of Prejudice*, Vol. 43, No. 5 (2009), 438.
93 Shekhovtsov, 'Apoliteic Music,' 429 & 450.
94 Sebastian Vehlken, *Zootechnologien: Eine Mediengeschichte der Schwarmforschung* (Zürich: diaphanes, 2012), 13. [My translation.]
95 *Vom Hirschkäfer zum Hakenkreuz*, dir. Madeleine Dewald & Oliver Lammert (Dock 43, 2002).
96 William H. Gass, 'The Order of Insects,' in *In the Heart of the Heart of the Country* (New York: Harper & Row, 1969), 167.
97 Roland Barthes, *Mythologies*, trans. Annette Lavers (New York: Farrar, Straus & Giroux, 1972), 120.
98 W.J.T. Mitchell, *What Do Pictures Want: The Lives and Loves of Images* (Chicago & London: The University of Chicago Press, 2005), 42.
99 *Vom Hirschkäfer zum Hakenkreuz*, dir. Madeleine Dewald & Oliver Lammert (Dock 43, 2002).
100 Sigmund Freud, 'On Fetishism,' in *The Complete Psychological Works of Sigmund Freud*, Vol. XXI, trans & ed. James Strachey (London: Hogarth & the Institute of Psychoanalysis, 1950), 152.
101 Freud, 'On Fetishism,' 154.
102 Steven Connor, 'Fascination, Skin & Screen,' *Critical Quarterly*, Vol. 40, No. 1 (April 1998), 12.
103 Connor, 'Fascination, Skin & Screen,' 12.
104 Jünger, *Subtile Jagden*, 22.
105 Jünger, *Subtile Jagden*, 34.
106 Connor, 'Fascination, Skin & Screen,' 9.
107 Jünger, *Subtile Jagden*, 34.
108 Jünger, *Subtile Jagden*, 77, 74, 85 & 91.
109 Jünger, *Typus, Name, Gestalt*, 88.
110 Jünger, *Typus, Name, Gestalt*, 123.
111 Freud, 'Medusa's Head,' in *Writings on Art and Literature*, trans. James Strachey (Stanford: Stanford University Press, 1997), 265.
112 Connor, 'Fascination, Skin & Screen,' 13.
113 Lacan, *Four Fundamental Concepts*, 180.

3 The Insectile *Informe*

H.P. Lovecraft and the Deliquescence of Form

In H.P. Lovecraft's 'The Whisperer in the Darkness' (1930), the 'deep things'—evidence of progressively colonising alien 'entities'—that the narrator Albert N. Wilmarth encounters in Vermont are relayed, especially, as sound phenomena, a phonic materiality that is rendered as insectile.[1] The unthinkable beings, whose formation occurs, in Graham Harman's words, according to a peculiar 'ontography'—a usage of language that is allusive and excessive, unable to crystallise the thing described into a coherent form—are crab-like and fungoid, but the sounds they emit nest a buzzing amidst, within, each and every word uttered. I am, here, interested in sound outwith—the Scottish preposition comes closest to capturing the interplay between outside and within—semantic content, represented as a 'thick droning voice,'[2] itself an outpost or archive of deep time. What I'm trying to do is to think through the concept of the insectile as sonic event and as it pertains to Lovecraft's deployment of that trope in 'Whisperer.' Here, but also elsewhere, in 'The Shunned House' (1937), for example, the insectile functions as a figuration of the outside—having to do with infinite time, what lies 'away outside,' beyond 'the last curved rim of space'[3]—and the formless, shapeless, nameless. Both of these aspects (the outside and the formless) are overwhelmingly heard, as well as felt, or sensed, as vibrations, buzzing voices that, recorded on a phonograph, travel in and out of apprehension. Technology, all in all, including the technology of writing, fails to capture these beings which, unlike Dracula, are not defeated by its systems.

This chapter positions itself against recent scholarship (easily) disposing of Lovecraft's racism, which in-forms—it internally forms, 'lives' inside—his 'Shoggothic Materialism'.[4] China Miéville has argued that

> all that kind of deep time, all that kind of deep *novum*, all that ecstatic collapse of the subject position [...] is predicated on master-race ideology; race hatred. So, in other words, the anti-humanism one finds so bracing in [Lovecraft] is an antihumanism predicated on murderous race hatred.[5]

DOI: 10.4324/9781003322757-3

At issue, here, is to insist on that 'deep' conceptual framework that, on the one hand, appears everywhere in Lovecraft's work and, on the other, has a tendency to disappear in criticism, especially if it is seeking to 'weird' philosophy through his fiction. Ben Woodard, for example, turns to Lovecraft to 'return' philosophy to the 'great outdoors,' Quentin Meillassoux's term for an anti-anthropocentric system of thought, but for all the speculative realists' provocative operations, their attempts to formulate a philosophy derived from Lovecraft, on 'the material of the external world' and without considering 'race hatred' as the absolutely coherent aspect of his work, require scrutiny and persistent opposition.[6] No matter what propositions arise in such readings, the apparent anti-, in- or posthumanism that these detect and promulgate as a result is toxic as well as unconvincing: the politics of this particular theoretical project must urgently be confronted.

The reason for its failure—a posthumanism in name only—is, precisely, that 'deep' entanglement with racism or, in other words, with the ideology of form, the particular ordering and valorising of a subject of 'wholesome stock'.[7] In contrast to critics who focus on a becoming-other in Lovecraft's stories, the basis for claims that credit him with writing about the 'impossibility of being a human in deep time,'[8] I contend that the 'human,' more specifically colonial man, is preserved as order-word beyond the 'affordances,'[9] the recurring patterns or 'signature' of the tentacle[10] or, as I propose below, of the insectile. While Lovecraft's stories have lent themselves to anti/in/posthumanist interpretations, these are, as mentioned above, unwilling to engage with the politics in-forming his writings, while they also dismiss the extent to which the other, should it eventually be approached with anything amounting to a welcome, is used to restore and maintain the form of the Same.

The examples given on this subject—Lovecraft as poster boy for the posthuman—frequently emerge from 'The Shadow over Innsmouth' (1936) or also 'At the Mountains of Madness' (1936), where the slippage of the I appears most compellingly. In 'The Shadow over Innsmouth,' the narrator, a fascinated subject—he finds himself arrested by the 'strange, unearthly splendour' of an alien tiara[11]—gradually discovers his lineage with fish-creatures, which arrive in dreams to draw him beneath the waters. The process unfolding here is sensuous, mesmerizing, exaltation replacing terror, fluidly, as sea change, a flow of desire. What is taking place is a decentring of the normative subject and yet this instance and possibility of becoming-other—the changing entity of the self effectively constitutes, as Miéville observes, an erotics[12]—are, on the one hand, obviously integrated into the Lovecraftian mythos and, on the other, displace the terror the narrator should 'properly' experience. This experience of the proper against the other, more specifically the Shoggoth—'intolerable,' 'all aout o'shape,'[13] according to the drunkard Zadok Allen—is 'propered' back to me, 'constant reader' perhaps (to refer to Stephen King's interpellation in every preface), or, at any rate, alert to the dangers that fascination poses to the phallic I.

The Shoggoth is a persistent 'figure' in Lovecraft's fiction; protoplasmic mass, it offends in its form or, rather, formlessness mocking the 'proper' subject, even if this subject seems, at first glance, as it does in 'At the Mountains of Madness,' other. As such, it operates in a tradition of 'viscous'[14] representation or order of ruin that is intimated at the end of 'Innsmouth' through the single word 'shoggoth' with its underwater resonance but also through the effect the creatures and their artefacts have on the entranced narrator. To lose one's head, a literal occurrence in 'Mountains,' is significant with respect to 'Innsmouth,' too, because of the phenomenon of a 'hypnotic order'[15] established in dream-states that, over time, abrade the narrator's conscious personality to make him more amenable to life beneath the water. Freud, by way of Gustave Le Bon, discusses this 'sacrifice' in *Group Psychology and the Analysis of the Ego* (1921), but the more famous instance of fascination and subsequent decapitation, read castration, occurs in 'Medusa's Head' (1922). The 'sight of Medusa's head makes the spectator stiff with terror,'[16] writes Freud, a reaction that is nowhere in evidence in 'Innsmouth' with its suggestion of the smooth, of the submerged I being smoothed out, eroded. Even daddy can't help the narrator 'secure' his place—notably through a job in an insurance office—but the contract drawn up is made with the reading subject, stiffening up to ward off evil. The erotics of the 'sea-deeps'[17] provokes that displaced defensive reaction which the narrator lacks but which is recovered in another I that, upright and vigilant, recoils, stiffly: it will not be castrated by, abraded into, the other.

'At the Mountains of Madness' provides 'salient'[18] details pertaining to the Shoggoth: subject-as-Shoggoth is slave subject and hence no subject at all but whose mimicry or mockery nonetheless defies, as it does in 'Innsmouth' and everywhere else in Lovecraft, the old order. In this instance, William Dyer, the first-person narrator, discovers that the 'Old Ones,' architects of fantastic cities in the Antarctic built by slave labour (i.e., the formless Shoggoths), are in fact kindred, their authority long usurped, their reign disastrously at an end:

> Poor devils! After all, they were not evil things of their kind. They were the men of another age and another order of being. [...] They had not been savages—for what indeed had they done? [...] [P]oor Old Ones! Scientists to the last—what had they done that we would not have done in their place? God, what intelligence and persistence! [...] Radiates, vegetables, monstrosities, star spawn—whatever they had been, they were men![19]

In the passage above, Dyer territorialises the other, the 'Old Ones,' as 'men of another age,' despite their physical appearance, 'scientists to the last,' and extends membership of an exclusive club, that is, the discourse of the human, to an Other who behaves as he does or 'would have done'. In other words, Dyer affirms the rights of those that are alike and who, consequently, are not subjects of difference but of his kind. He thereby bestows legitimacy onto their order and integrates them into his political community: 'whatever they had been, they were men!' This neutralisation of difference complicates arguments, to say the least, about Lovecraft's anti/in/posthumanism; the logic at work, rather, upholds 'man' as standard-bearer by which to assess the other, granted recognition in relation to a political norm. This chapter contests the basis on which these arguments are made, and its contention is the following: the form of the 'proper' subject remains that of 'man,' an organisation of form and of the proper that articulates and must secure itself against a formless other.

This is not to say that Lovecraft's work is homogenous, although my intention is not to demonstrate its possibilities of rupture or transgression; according to Derrida, these always exist in a text, even in the most phallogocentric ones, which might produce 'paradoxical effects'.[20] Lovecraft, regardless, has his advocates, and, more than anything, the impression persists that recent scholarship insists too much on, or hallucinates, disturbances in the order of the 'human,' whose precarity exists to the extent that it has to be protected. A community of the 'we' either is silently assumed or, as in 'The Case of Charles Dexter Ward' (1943), is explicitly and 'dutifully' taking action; this concept of duty motivates 'guiding group[s] of eminent men' standing guard and raiding the scene of the other.[21] A 'we' similarly recurs in Harman's *Weird Realism* (2012)—the epicentre, if you wish, of the present critique—which Ezra Claverie, in his book review of Harman's study, identifies as the perspective of the coloniser, the white supremacist harmoniously aligned with Lovecraft's mythos.[22] There are aspects of *Weird Realism* that are astute, not least of which is Harman's description of Lovecraft's style in terms of 'literary cubism,'[23] delivering objects from multiple viewpoints at once. This practice of writing—it is compound, like an insect's eye—yields new, formless, arrangements, split and unreconciled in their dimensions, partly withdrawn and/or overwhelmingly, disastrously, there. Words thus function like black holes deforming everything around them (like the usual qualities associated with a certain term).[24] Indeed, Lovecraft's writing is morphological, interested in forms and un-forming, but the incentive driving this process is the privileging of form and, more precisely, the modality of the colonial man/white settler, the measure of all things or of all life. A writing style that disjoints, creates rifts, breaks things apart does not necessarily come from, or engender, a similarly 'deconstructive' movement, meaning an operation that calls into question structure and form. On the contrary, Lovecraft's ontography, rather than being an anti-structuralist gesture, is resolutely occupied with preserving systems of formalisation, including the logic of racialisation.

To understand Lovecraft in a tradition, or direction, exposed to the inhuman, insistent on the 'value' of horror fiction as philosophical exercise/element—because why waste your time looking at something unworthy of attention, hence Harman's effort to give it proper form—doesn't preclude being unresponsive to the other. Questions of value are at the heart of this chapter, as they are to Harman's book (a rescue mission; a defence of Lovecraft's literary merits) and to inquiries about the *informe*, that which, according to Georges Bataille, 'does not, in any sense whatever, possess rights, and everywhere gets crushed'.[25] The insectile *informe*, as the title announces, sets the course of this chapter; linked to an insect's metamorphosis, the *informe* is figuration, 'living map'[26] of a dynamic, constantly changing subject different from itself, whose identity is not repose but transformation. This dynamism is often imagined as deliquescence, a term which is borrowed from mycology and which suggests itself through the 'fungous life' in 'Whisperer' but also throughout the rest of Lovecraft's oeuvre: the fungus's putridity corresponds to his creatures' 'damnable approach to form'.[27] Deliquescence, however, not only stands in relation to a single, 'degrading' body but further calls up the (im)materiality of swarming multiplicities, which, according to Maurice Maeterlink, observing ants, '[present] the appearance of a liquid in ebullition'.[28]

Form and the deliquescence of form: these events—or, rather, formations, deformations, formatting, and transformations, to refer to Sebastian Vehlken—give shape to this chapter, which proceeds by way of a close encounter, of a kind, with buzzing voices in 'Whisperer.' In Miéville's analysis of the weird, the insectile is grafted onto the tentacular, which is privileged by Miéville (and Roger Luckhurst) as sign of displacement of the so-called human; the tentacular, however, misses the dimension of sound. The insectile is aural, not necessarily a visual phenomenon: it is cacophony, noise, murmur, buzz, 'indistinguishable,' according to Eugene Thacker, 'from the very elemental properties of [...] storms and whirlwinds.'[29] Miéville defines the creatures of the weird as 'indefinable and formless,' before exemplifying this impossibility of description and approach in a notation: 'and/or' or also 'and/or and and/or or'.[30] The notation itself demands a response: the forward slashes defer, are two- or multi-faced, each facet uneasily co-existing, the symbol at once separating force and a barrier to be overcome, inviting, in fact, its negotiation. 'Disproportionately insectile/cephalopodic,' the creatures of the weird, Miéville writes, are 'without mythic resonance,'[31] that is, unable to sound across a culture to provide it with a structuring form (of belief, explanation, etc.). Wedged between the insectile and the cephalopodic, the mark, however appropriate to Lovecraft's ontography, nonetheless makes it easy to gloss over the 'phenomenally complex' aspect of the swarm, which is 'affectual before [it is] accountable'.[32] There is, hence, a point to be made about pausing, for a moment, to consider the insectile, so as to concentrate on the realm that we might otherwise glide over, on the way to somewhere or something else.

Before becoming (and not even necessarily so) source, the insectile is environment. In its immateriality, it is without form, pertaining, as such, to another aspect of the *informe*, namely what Yve-Alain Bois and Rosalind Krauss consider to be 'pulsation.' A sort of beat which disrupts or 'punctures' the apparent closedness of forms, a 'unified visual field,' pulsation agitates, for example, Man Ray's photographs, though Bois and Krauss also mention Freud's *Beyond the Pleasure Principle* (1920), the 'rhythm of shock,' the urgency of pulse as repetition.[33] It behaves as a largely visual, rather than aural, category in their work, if simply because their analysis focuses on art, but pulse or pulsation, as Thacker has shown, is also swarm, a permeating aurality. The *informe*, here, then, refers to the pulsation of a resonant, sonorous materiality, whose theoretical framework is assembled from research into sound studies, the voice, media technologies, the impressions of sound-objects and their 'liquid instability,' the vibrations always in excess of the sources that emit them.[34] As Isabella van Elferen has shown, sound is a 'privileged metaphor' for Lovecraft's 'paradoxical materialism' because of its 'uneasy fit between ontology and phenomenology'—engendered, as it is, by instruments and voices but also obviously generating affective experiences—as well as between materiality and immateriality.[35] While van Elferen uses the matter of sound to point to the serious divergences between Lovecraft and speculative realist philosophy, I stay attentive to the flesh of sound, the auditive texture of voices like a bee's, in order to demonstrate Lovecraft's orientation toward form. The diffusive insectile buzzing, in and beyond 'Whisperer,' endeavours to represent that which is unrepresentable, the 'thought from the outside,' in Michel Foucault's terms.[36] Rendered in a language whose vibrations, its timbre, performs—or, rather, in which resonates—that pure outside, the insectile exists as evocation or pulsation of the formless.

What is to come is an engagement with the valorisation of form (i.e., the good form), against that which is deemed formless/deliquescent/insectile. The first part of the chapter travels through the methodology on the materiality of the voice and thereby lends an ear to the flesh of the buzz, while the second is concerned with the assemblage and notation of the good form in Lovecraft's 'Whisperer' by being alert to punctuation as indicative of the voice that is expelled from the order of writing. A project about morphology, it attends to sound, to the correlations between sound and form, and addresses the obfuscation at the heart of recent scholarship on Lovecraft, whose particular iteration of anti/in/posthumanism can't finally offer any sustained or substantial critique of the systems of thought sustaining the violent political domination of the 'human,' a.k.a. colonial man. Instead, such perspectives reconcile posthumanism with racialisation, with the construct of 'man' so apt at reconstituting itself at every turn. Rather than a manifestation of 'outside thought,' in the sense of working to decolonise the concept of the 'human,' this mode of thinking actively neglects the production of form (colonial man) and formlessness (the insectile) in Lovecraft's writing, which, deforming, unforming, nonetheless upholds the value-form of the master–subject.

The Flesh-voice of the z

It is the voice (as well as handwriting, which becomes impossible because distorted) that gives away otherness in Lovecraft's fiction, even if, on the face of it, appearances are more or less kept up. In 'The Shadow Out of Time' (1936), Nathaniel Wingate Peaslee, professor in Political Economy at Miskatonic University, descended of 'wholesome old Haverhill stock,' returns to occupy his own body after an abduction lasting for over five years and finds that his 'speech seemed awkward and foreign':

> I used my vocal chords clumsily and gropingly, and my diction had a curiously stilted quality, as if I had laboriously learned the English language from books. The pronunciation was barbarously alien, whilst the idiom seemed to include both scraps of curious archaism and expressions of a wholly incomprehensible case. [...] Something in my aspect and speech seemed to excite vague fears and aversions in every one I met, as if were a being infinitely removed from all that is normal and healthful.[37]

In 'The Case of Charles Dexter Ward,' Joseph Curwen, the dodgy ancestor, adopts the speech of a 'learned and cultivated Englishman' but whose mimicry hints at 'sinister undercurrent[s]'.[38] In both these cases, language is something improperly acquired, as if it wasn't theirs to possess, or their improper bodies were barred from being claimed by, or associated with, the English language. In 'Whisperer,' too, 'buzzing voices' speak 'in imitation of human speech'; these are not 'well-built' voices but 'animal noises,' a speech 'decayed,' misshapen, unable to really or properly be formed into words.[39] There is an intolerable excess to these speaking voices, all the more unbearable because illegitimate, 'laboriously' appropriated rather than 'naturally' pertaining to the subject and wider order in question. This discrepancy, between 'nature' and labour, as Homi Bhabha shows in *The Location of Culture* (2004), threatens the authoritative discourse of colonial man, an 'empty' subject or, in Lauren Berlant's words, abstract citizen.[40] Abstract personhood is not, as the term already implies, bound to a living body; if a body appears—a fat, black, queer, differently abled body, say, or a crustacean or Shoggothic body—it is surplus, a 'bad form' whose corporeality also lodges in the voice.

There is a weight, a viscous density, to the shaping labour of the other's language—even the term 'body' bestows too much form—whose messages can't be free of its medium, that is, the voice, emerging from organs 'unmistakably alien to this world'.[41] The materiality of the voice—residing, for one, in the 'fiendish' presence of punctuation marks brutalising the writing, the separate spheres of each letter—is impossible to be passed over and signifies a perversion of form even more 'blasphemous' because of the 'impersonal precision and deliberation' informing the mimicry.[42] Soulless, mechanical, insectile, this voice destabilises colonial man/history/authority through its fleshiness, that which, with reference to Alexander Weheliye, is extraneous according to the logic of the political domination of the other, sustaining and activating an 'atrocity of the flesh'.[43] Colonial man has no flesh, just as his voice isn't marked by a surplus viscosity; racialisation is excessive fleshiness, made apparent, on the pages of Lovecraft's story, through the way messages are forced through an intolerable body. The buzzing is the result of this trajectory, the purity of a language sullied by organs unfit to pronounce it and all the more monstrous for its 'cultivation'.

Jean-Luc Nancy's *À L'Écoute* (2002) draws the ear toward the enfleshment of the voice or its timbre, what Roland Barthes, acknowledging Julia Kristeva's work, calls the 'geno-song,'[44] that which is remaindered from the speaking voice as source of meaning or semantic order. Rather than valorising the message, Nancy asks us to listen to sonority: he uses the expression *tendre l'oreille*,[45] preserved in the English translation of the book, that is, to stretch the ear, mobilise it, and attend to its travels. 'To be listening is always to be on the edge of meaning,' he writes, considering sound as edge, 'open depths,'[46] an outside in which the subject is immersed and which resonates within. What is remarkable is that Nancy eschews a vertically held perspective (the all-seeing I) for a position that renders the listening subject as resonant womb or belly, a 'hollow column, over which skin is stretched'.[47]

A philosophy of listening to approach Lovecraft's 'The Whisperer in the Darkness,' the materiality of his buzzzzzing voices: the way the z, like skin, can stretch or hummmm. In an interlude, Nancy plays with the word *mot*, the noise of m, to murmur, to mutter, to become mute, to pour out, via the German *münden*, also incubating the mouth.[48] A buzzzzzing: the vocal folds are tense and vibrate together, their vibrations sensed as material, the distended z vibrating at the roof of the mouth, the tongue lightly touching its ridge. The airflow is restricted but not stopped; a small opening allows it to pass, hitting the back of the teeth. In the 'cave of the mouth,'[49] a nervous entity nests: not yet, no longer, a word, it is a disturbance, a bug hovering, writhing, within. The buzzing of the consonant m articulates no voice, writes Nancy, a compelling observation though not strictly true. The buzz is voice as excessive material object, as 'vocality,' to use Paul Zumthor's term, which Adriana Caravero uses to discuss the dimensions of the voice as far larger than speech.[50] Caravero proposes that logocentrism denies the voice its range, its 'seductive and quasi-animal' reverberations.[51] The voice of the othered is frequently dismissed as noise, hysterical babble, and so on, but the logocentric command must render its own voice as pure meaning: beautiful, clean, uncluttered information machine.

Mladen Dólar has analysed the 'linguistics of the voice,' referring to Aristophanes' hiccups in Plato's *Symposium*, irruptions usually understood as 'recalcitrant to meaning,' like coughing, or any other 'soulless,' automatic, involuntary interruption of speech.[52] Dólar, however, shows that such interferences—parasites, in Michel Serres's vocabulary—while not linguistic, do not simply lie outside of linguistic structure either. He considers them, in 'their very absence of articulation'—what Nancy might have gestured toward with his comment concerning voicelessness—as the embodiment or corporeality of linguistic structure as such.[53] Elsewhere in his study, he describes the voice as 'plus-de-corps,' a double entendre that, at the same time, hides the body (plus de corps = no body left) and increases its mass (plus de corps = supplementing the body with yet more flesh).[54] The incorporation of the voice, though it never really belongs to the body it emanates from, is evident in the sonorous body Nancy posits, echoing Derrida's alertness, his stretched ear, to 'the irreducible openness of the inside,'[55] making any absolute inside, closed off from the other, impossible.

The exact location of the voice always remains in doubt, even if we can describe exactly where and how sounds are formed. Z is a voiced alveopalatal sibilant fricative, occurring because a certain number of conditions are fulfilled for its emergence and lingering vibrations. In his book about sound and avant-garde art, Douglas Kahn notes that the voice inhabits bodies differently:

> modern Western culture typically locates the dominant operations of the embodied voice above the collarbone, attracted toward the head by the pull of the fusion of thought with speech and by an unconscious that serves as a proxy for the rest of the body. Other cultures place the operations of the voice throughout the body, and some place them primarily below the collarbone and symbolise voice through an array of objects, economies, and forces both inside the body and well outside of it.[56]

In this particular chapter, Kahn examines what he calls 'meat-voices,' a voice spread throughout the entire body, in operation, for example, in William Burroughs's work, in the word *schlupp*—expressing hunger or desire—in which the sounds of the body can be abundantly heard. It takes moisture to pronounce this word, slushing inside, in the wet region of the mouth. It is consequently as if a word, made flesh, released spores, starting an autodigestive process: this affective quality pertains to the basely material, the entropic;[57] in other words, to the *informe*. It does not so much form an image as deform, degrade or transform it; like slime, which Sartre analyses in *Being and Nothingness* (1943), its softness, its 'inexpressible materiality,'[58] is indicative of formlessness, the dissolution of representation. In the 'event' of the Lovecraftian buzzing voice—'metallic, lifeless,' with its 'inflectionless, expressionless scraping and rattling, and its impersonal precision and deliberation'[59]—it similarly calls up a horror of that which cannot become, and explodes, form.

The flesh-voice of the buzz—flesh so as to retain, in a first instance, the association with Weheliye's work—rather than being integrated, is a distributed, disaggregated voice, in that it can be located across a number of bodies. Its materiality is vibrational; it is a swarming, a noise indicative of a 'semi-being,' as Leibniz calls multiplicities.[60] In *Genesis* (1982), Serres suggests that multiplicities—a pack, a swarm, a herd—are 'a bit viscous perhaps,' hardly objects (he likens them to a lake in mist, a white plain),[61] though they are frequently produced as un- or anti-aesthetic thing and, in different contexts, as specific form or figure of knowledge, as Sebastian Vehlken shows in *Zootechnologien* (2012). Form, as Vehlken notes, is not an ontological condition but a historical/political phenomenon; a swarm, as 'epistemic object,'[62] does not remain statically the same but is rendered in and according to particular discourse networks across time. In horror fiction, however, it tends to fulfil the same function, figuration of the total other linked, time and again in Lovecraft's story, with the 'untenanted' or with perverse tenancy: swarms occupy hills, which nobody visits, 'tenantless mountains,'[63] the buzz of the voice itself suggesting further tenantlessness, that which has no soul. If the flesh, in its surplus viscosity, 'living fungi,'[64] as it were, designates an absence from a subject position—'before the "body," there is the "flesh"', writes Hortense Spillers—the buzz is one expression of 'that zero degree of social conceptualisation' afforded to 'formless' beings.[65] It is also the voice of *zoē*, the lack of *eidos*, of something without form, head or face, of a thing that is 'dangerously close to the arcana [the mystery or secret] of basic entity [...] [transcending] form and force and symmetry'.[66]

The holy trinity of value, that is, form and force and symmetry, defines the discourse of absence with respect to basic entities, coded as a type of amoeboid life—without form, like the horror-swarm—as subjectless assemblage with no speech of its own. In the buzz resonates that elimination of form, what either can't be formed into words or, at best, nests within them, threatening their 'voiceless,' fleshless, integrity. The extraneous noise, that voice like a bee's, corresponds to the real of a 'vast outside,'[67] swelling beyond the background to the good form to dissolve it. There is no 'logos without noise,'[68] which forms the underside (the outer limits) of information: noise is the material of form, from which form is shaped and which simultaneously threatens its order.

The Good Form

Up until now, the more or less silent but nonetheless foundational influence to this chapter is Seb Franklin's 'The Context of Forms,' in which he analyses form and formlessness in light of 'the complex, ever-shifting knot of settler colonialism, exploitation, and immiseration otherwise known as capitalism'.[69] His study, flanked on either side by references to Lovecraft, examines the dynamics of formalisation entangled with commodity production—of bodies (human; non-human) and territories—always reliant on surplus formlessness as something that awaits being made productive. Capitalism moulds things into forms, contorts the shape of the worker and confers value on that which has been formed, made useful, yet while demanding and fetishizing form, it also depends on that which still, temporarily, lies outside its processes of accumulation. Lovecraft functions as moment of entry to, and endpoint of, an investigation into the socio-historical conditions of formal assignation at the hands of capital. In other words, Franklin is interested in subsumption, formal and real, and the 'originary and ongoing violence' which determines form as well as formlessness, both of which are integral to racial capitalist procedures and social formations.[70]

Lovecraft's commitment to form in relation to an aspect that immediately refers to, or, if you will, succeeds, Franklin's work exemplifies the relationship, in 'Whisperer,' between formalisation, settlement and profit on the one side and formlessness, as excrescence, on the other:

> Most people simply knew that certain hilly regions were considered as highly unhealthy, unprofitable and generally unlucky to live in, and that the farther one kept from them the better off one usually was. In time the ruts of custom and economic interest became so deeply cut in approved places that there was no longer any reason for going outside them, and the haunted hills were left deserted by accident rather than by design.[71]

The misrecognition of accident and design is itself already significant, if accident is taken to mean something that 'naturally' occurs, though words like 'approved' or 'economic interest' are indicative of, precisely, design. An approval is an attestation of authority, and it is suggestive of form, a marking of territorial boundaries, coded as healthy and productive, whereas the 'outside' is to be distrusted, regardless/despite of its significance in sustaining and setting off the inside as all the more desirable. Here is a gaze that claims and abjects, a politics that is everywhere apparent in Lovecraft's writings, his fiction and letters, pitting '[men] of character, education and intelligence' against 'misshapen outcasts' in 'hidden and unwholesome [tenancies]'.[72] Though 'destinies' of the human and the non/in-human might be intertwined—'pits of primal life' trickling down into 'our own' pools[73]—the logic at work, so resonant with mechanisms of racialisation, is directed against an enemy within, an inside that must be expelled. The point is not so much that a stable separation between inside and outside, or form and formlessness, is impossible but that the former has to be safeguarded against the latter. The rights of the settler and, by extension, the identity of the 'human' are, as such, troubled only in the sense that they require an even deeper commitment to their maintenance, while the system of thought that grants form stays unchallenged.

It is, then, surprising, not to say irresponsible, that Lovecraft is claimed by a philosophical system apparently attentive to the non/in-human, bearing in mind how insistently he corroborates form, in the face of its deliquescence, and the form-giving structures of racial capitalism. Form might imply vision—and vision, in Western metaphysics, posits a rational, self-identical subject in command—but it has a sonic dimension. Form is audible as well as visual, though sound is even more readily linked to dissolution or, at the very least, to diffusion. In *À L'Écoute*, Nancy prehends the sonic as existing beyond form, in that it might evoke a form but can't be expressed (that is, contained) as such:

> [Sound] does not dissolve [form], but rather enlarges it; it gives it an amplitude, a density, and a vibration or an undulation whose outline never does anything but approach. The visual persists until its disappearance; the sonorous appears and fades away into its permanence.[74]

What arises in/as sound is something that can travel, stretch; it seems by definition to be formless, though it isn't. Harmony is form—music is an endowment of noise with form, according to Jacques Attali—emerging from a background noise that is 'limitless, continuous, unending, unchanging'.[75] Sound liquidates form in the sense that it passes the bounds of visual form. A sound is, hence, of uncertain ontological status, just as it is 'placed under the sign of a fall,'[76] as language's remainder, the excess of the voice, unnecessary for, even disruptive to, its messages.

The intertwining of materiality (the source of a sound) and immateriality (a sound's distribution; its affect), and the reason, remembering van Elferen's argument, that sound features so prominently in Lovecraft's mythos, gives it its function with respect to form. Serres likens sound to a phenomenon like radiation, settling in subjects and things, moving through flesh and walls, 'bounding, abounding, unbounding' space;[77] its coextension with space also comes with the faculty or the incapacity of filtering it out. Sound can, then, have form, or rather can be arranged into form, and undo form, often at the same time: if a source is visible, can be isolated, it doesn't mean that sound can be captured as belonging, being proper, to that single source. Hearing as such is not necessarily something that occurs before formalisation but is in fact trained to respond to it. The nominal good is, yet again, form, which affects our ability and/or willingness to listen to that which interferes with said good. 'I hear without clear frontiers,' observes Serres, continuing that the 'ear knows how to lose track,'[78] but 'Whisperer,' although *à l'écoute*, is case in point of hearing as obedience, attending exclusively to that which has already been coded. The ear, in these circumstances, is an organ receptive to given orders/forms and standing in stark contrast to 'malformations,' the kinds of sounds that are made by matter utterly remote to a particular understanding of the good form.

In this vein, and in a lengthy passage, the narrator in 'Whisperer' describes his first and traumatically retained point of contact with 'outer' voices. Initially, the encounter (not first-hand) is relayed via mailed transcript, lost but etched into memory, then by phonograph recording received from his correspondent Henry Wentworth Akeley (note his middle name, referring to Governor Wentworth, dispensing 'colonial grants' to settlers in what became the state of Vermont):

> To me, with my first-hand impression of the actual sounds and with my knowledge of the background and surrounding circumstances, the voice was a monstrous thing. It swiftly followed the human voice in ritualistic response, but in my imagination it was a morbid echo wringing its way across unimaginable abysses from unimaginable outer hells. It is more than two years now since I last ran off that blasphemous waxen cylinder: but at this moment, and at all other moments, I can still hear that feeble, fiendish buzzing as it reached me for the first time.
>
> [...]
>
> But though the voice is always in my ears, I have not even yet been able to analyse it well enough for a graphic description. It was like the drone of some loathsome, gigantic insect ponderously shaped into the articulate speech of an alien species, and I am perfectly certain that the organs producing it can have no resemblance to the vocal organs of man, or indeed to those of any of the other mammalia [sic]. There were singularities in timbre, range and overtones, which placed this phenomenon wholly outside the sphere of humanity and earth-life.[79]

In the story, Lovecraft repeatedly mentions the disruption caused to channels of communication, letters lost, phonograph records destroyed, wires going dead, evident references, as mentioned earlier, to *Dracula* (1897): in this case, however, the vampiric systems of telecommunications are not equal to the task. The phonograph, as Friedrich Kittler tells us, seizes the real, the excesses or residue of the voice, everything whispered, rattled, all that which is excluded in the discrete notations of the alphabet.[80] It is a technology of noise or of 'waste'—the Viennese psychiatrist Erwin Stransky used the phonograph to register the 'uninterrupted, indiscriminate'[81] utterances of his subjects—though it was initially developed as 'archival apparatus' to preserve the 'exemplary words' of political leaders, in other words, the discourse of power, rather than the babblings of the unconscious.[82] Employed, in 'Whisperer,' to make analysis possible, the mechanism succeeds only in entrenching the trace or trauma of pure difference, its spiral groove recording and reproducing, in memory, the sound waves of an 'outward logic'.[83] The outside is, in this way, reterritorialised as something which is inside but cannot, at the same time, be admitted.

It is not as if the network of technique, at any rate disabled, managed to process buzzing as object or function of an organism that can somehow enter representation. Form, or form-making, fails here, but not exclusively because machines, and the technology of language, fail. A discourse network, needless to say, is ideological, the failure, consequently, also one of the imagination. The recordings follow an already grooved loop, in that the impressions they make are guided by a colonially organised idea of what form is, a program that is apparent when looking at how these recorded iterations are relayed in the story. Punctuation is key here: 'marks of oral delivery,' they are usually 'friendly spirits,' according to Theodor Adorno, bound up with the 'musical form,' the schema of tonality.[84] The remembered transcript of the phonograph record, preceding Wilmarth's reaction shown in the previously cited passage, however, is evidence of being unsettled, displaced. Punctuation marks—largely ellipses but also exclamation points and square brackets, either filled with descriptions of the 'outside' voices in italics or supplementing garbled words—are 'bodiless' indicators that, here, do not resemble music (a culturally sanctioned form) but disorder.[85] Indicative of an 'interplay that takes place in the interior of language'[86]—'good English grammar,'[87] according to Wilmarth, disrupted and undone by parasites—punctuation marks disturb the good form from within. If the phonograph hears everything, the transcript (and its engraving in memory) records chaos through the 'noise' of frequent elliptical interludes, signalling incomprehension, a trailing off, unmoored, into infinitude. The 'silent cymbal clashes' of exclamation points, the 'enclave[s]' of the parentheses, corrupt the 'integrity' of the sentence and of the 'linguistic form'.[88] For Adorno, the 'micrological power' of punctuation—a comment about Proust's use of punctuation marks but repurposed, here, to talk more generally about their potential—lies in the silent ways it approximates writing to the voice, to the sound that writing suppresses.[89] The impact of the excessive use of punctuation marks is degeneration, an assault on the symbolic order by the terror of the voice of the other.

Defined not only through its, or his, capacity to use language—the default position by which to assert 'human' supremacy over the animal other—but through the ability to use (particularly the English) language well, the 'human' is determined as, and with respect to, English colonial man (Akeley; Wilmarth). Transcript and phonograph record stage a conversation between outer voice and recruit or 'reformed' subject,[90] speaking in a 'mellow, educated voice which seemed vaguely Bostonian in accent'.[91] This reformed subject—a Mister Noyes, onomatopoeically referring to what is provoked; later also Akeley—is an ambassador of sorts; both his voice of privilege and the voice of the other, exerting an 'almost paralysing fascination,' unsettle (un)familiar sounds to the extent that they reveal their repressed source, namely that all language is mimicry:

> At times it seemed as if he [Akeley, by this point possessed] were pumping me to see what I knew of the monstrous secrets of the place, and with every fresh utterance that vague, teasing, baffling familiarity in his voice increased. It was not an ordinary or healthy familiarity despite the thoroughly wholesome and cultivated nature of the voice. I somehow linked it with forgotten nightmares, and felt that I might go mad if I recognised it.[92]

The voice is that which arrives from outside to take up residence within, but the result of this not-quite recognition—exiled from the ego—is less a dethroning of the properly colonial man than a consolidation of his power. Even though the story, in a way a retelling of the Medusa myth, appears to demonstrate how easily man is de-phallusised, pumped of his *Gestalt* and *bios*, it obsessively and aggressively seeks to maintain what Wynter calls the 'coloniality of Being,' over-representing colonial man as the 'human' per se.[93] Considering that the 'bare' matter of the brain is the only organic residue left of this subject and that his/its voice, too, is losing form—Akeley, towards the end of the story, speaks in a monologue full of dashes, openings toward the 'gulfs of space and time' now heard in the voice—the assault on the 'human' is total, unrelenting.[94] And yet this figure persists: his form upheld as centre of all reference and the appropriate instantiation of the 'human,' because the narrative in all instances prioritises the 'wholesomeness' of form over its deliquescent other: indescribable, fungoid flesh, the pulsations of the insectile.

In light, then, of the analysis above, the mythic resonance that Miéville fails to see with respect to the formless occurs elsewhere, that is, in the construct of form, privileged over and against the insectile. This construct or myth, both originary and ongoing, is the ultraviolent logic of organisation that defers to colonial man, articulating the following equation: colonial man = eternal good form. What resonates not only throughout Lovecraft's work, but also across the scholarship establishing and legitimising him as a writer stretching his ear towards the non/in/anti-human, is the discourse of the valorisation of form: 'man'—the ultimate apparition of form—emboldened against the murmur, the buzzzzzzzzz of a background without form.

Notes

1 H.P. Lovecraft, 'The Whisperer in the Darkness' in *The Whisperer in the Darkness and Other Stories* (London: Wordsworth, 2007), 217.
2 Lovecraft, 'The Whisperer in the Darkness,' 237.

3 Lovecraft, 'The Whisperer in the Darkness,' 241.
4 Ben Woodard, 'Mad Speculation and Absolute Inhumanism: Lovecraft, Ligotti and the Weirding of Philosophy,' *Continent*, Vol. 1, No. 1 (2011), 4.
5 China Miéville & Jeffrey Andrew Weinstock, 'Afterword: Interview with China Miéville' in *The Age of Lovecraft*, ed. Carl H. Sederholm & Jeffrey Andrew Weinstock (Minneapolis: University of Minnesota Press, 2016), 241.
6 Woodard, 'Mad Speculation and Absolute Inhumanism,' 9 & 3.
7 H.P. Lovecraft, 'The Shadow out of Time' in *The Lurking Fear and Other Stories* (London: Panther Books, 1964), 157.
8 Miéville & Weinstock, 'Afterword: Interview with China Miéville,' 236.
9 Caroline Levine, *Forms: Whole, Rhythm, Hierarchy, Network* (Princeton & Oxford: Princeton University Press, 2015), 29.
10 Roger Luckhurst, 'The Weird: A Disorientation,' *Textual Practice*, Vol. 31, No. 6 (2017), 1045.
11 H.P. Lovecraft, 'The Shadow over Innsmouth,' in *The Lurking Fear and Other Stories* (London: Panther Books, 1964), 109.
12 Miéville & Weinstock, 'Afterword: Interview with China Miéville,' 235.
13 Lovecraft, 'The Shadow over Innsmouth,' 121 & 132.
14 H.P. Lovecraft, 'At the Mountains of Madness' in *At the Mountains of Madness & Other Novels* (London: Gollancz, 1971), 89.
15 Sigmund Freud, *Group Psychology and the Analysis of the Ego*, trans. James Strachey (London: The Hogarth Press, 1949), 10.
16 Sigmund Freud, 'Medusa's Head,' in *Writings on Art and Literature*, trans. James Strachey (Stanford: Stanford University Press, 1997), 273.
17 Lovecraft, 'The Shadow over Innsmouth,' 155.
18 Lovecraft, 'At the Mountains of Madness,' 57.
19 Lovecraft, 'At the Mountains of Madness,' 90.
20 Jacques Derrida, *Acts of Literature*, ed. Derek Attridge (New York & London: Routledge, 1992), 50.
21 H.P. Lovecraft, 'The Case of Charles Dexter Ward' in *At the Mountains of Madness & Other Novels* (London: Gollancz, 1971), 132.
22 Ezra Claverie, 'Weird Interpellation,' *new formations*, Vol. 84, Autumn 2015, 264.
23 Graham Harman, *Weird Realism: Lovecraft and Philosophy* (Winchester & Washington: Zero Books, 2012), 234.
24 Harman, *Weird Realism*, 239.
25 Georges Bataille, *Encyclopaedia Acephalica: Comprising the Critical Dictionary & Related Texts*, trans. Iain White (London: Atlas Press, 1995), 51–52.
26 Rosi Braidotti, *Metamorphoses: Towards a Materialist Theory of Becoming* (London: Polity, 2002), 2.
27 Lovecraft, 'The Whisperer in the Darkness,' 261 and 'The Shunned House' in *At the Mountains of Madness & Other Novels* (London: Gollancz, 1971), 243.
28 Maurice Maeterlink, *The Life of the Ant*, trans. Bernard Miall (London: George Allen and Unwin, Ltd., 1941), 34.
29 Eugene Thacker, 'Pulse Demons,' *The Funambulist*, 13 Jan 2013. Accessible via: https://thefunambulist.net/philosophy/philosophy-pulse-demons-by-eugene-thacker [Accessed 29 April 2022].
30 China Miéville, 'M.R. James and the Quantum Vampire: Weird; Hauntological: Versus and/or and and/or or?,' *Weird Fiction Review*, 29 November 2011. Accessible via: http://weirdfictionreview.com/2011/11/m-r-james-and-the-quantum-vampire-by-china-mieville/ [Accessed 29 April 2022].

31 Miéville, 'M.R. James and the Quantum Vampire.'
32 Thacker, 'Pulse Demons.'
33 Yve-Alain Bois & Rosalind E. Krauss, *Formless: A User's Guide* (New York: Zone Books, 1997), 31 & 162.
34 Steve Goodman, 'The Ontology of Vibrational Force,' *The Sound Studies Reader*, ed. Jonathan Sterne (London & New York: Routledge, 2012), 71.
35 Isabella van Elferen, 'HYPER-CACOPHONY: Lovecraft, Speculative Realism, and Sonic Materialism,' *The Age of Lovecraft*, ed. Carl H. Sederholm & Jeffrey Andrew Weinstock (Minneapolis: University of Minnesota Press, 2016), 93.
36 Michel Foucault, 'Maurice Blanchot: The Thought from the Outside,' in *Foucault/Blanchot*, trans. Brian Massumi (New York: Zone Books, 1987), 16.
37 H.P. Lovecraft, 'The Shadow out of Time,' 157 & 159–159.
38 Lovecraft, 'The Case of Charles Dexter Ward,' 113.
39 Lovecraft, 'The Whisperer in the Darkness,' 220 & 228.
40 Homi K. Bhabha, *The Location of Culture* (London: Routledge, 2004), 123 & Lauren Berlant, *The Queen of America Goes to Washington City: Essays on Sex and Citizenship* (Durham & London: Duke University Press, 1997), 18.
41 Lovecraft, 'The Whisperer in the Darkness,' 230.
42 Lovecraft, 'The Whisperer in the Darkness,' 235.
43 Alexander G. Weheliye, *Habeas Viscus: Racialising Assemblages, Biopolitics, and Black Feminist Theories of the Human* (Durham & London: Duke University Press, 2014), 2.
44 Roland Barthes, 'The Grain of the Voice,' in *The Sound Studies Reader*, ed. Jonathan Sterne (London & New York: Routledge, 2012), 506.
45 Jean-Luc Nancy, *Listening*, trans. Charlotte Mandell (New York: Fordham University Press, 2007), 18.
46 Nancy, *Listening*, 21 & 16.
47 Nancy, *Listening*, 77.
48 Nancy, *Listening*, 46–47.
49 Nancy, *Listening*, 48.
50 Adriana Cavarero, 'Multiple Voices,' *The Sound Studies Reader*, ed. Jonathan Sterne (London & New York: Routledge, 2012), 528 & 529.
51 Cavarero, 'Multiple Voices,' 529.
52 Mladen Dólar, *A Voice and Nothing More* (Cambridge, MA & London: MIT Press, 2008), 48 & 73.
53 Dólar, *A Voice and Nothing More*, 97.
54 Dólar, *A Voice and Nothing More*, 200.
55 Jacques Derrida, 'The Voice that Keeps Silence,' *The Sound Studies Reader*, ed. Jonathan Sterne (London & New York: Routledge, 2012), 502.
56 Douglas Kahn, *Noise, Water, Meat: A History of Sound in the Arts* (Cambridge, MA & London: MIT Press, 2001), 290.
57 Bois & Krauss, *Formless*, 26.
58 Jean-Paul Sartre, *Being and Nothingness*, trans. Hazel E. Barnes (New York, London, Toronto, Sydney & Singapore: Washington Square Press, 1992), 606.
59 Lovecraft, 'The Whisperer in the Darkness,' 271.
60 Michel Serres, *Genesis*, trans. Geneviève James & James Nielson (Ann Arbor: University of Michigan Press, 1999), 2.
61 Serres, *Genesis*, 5.
62 Sebastian Vehlken, *Zootechnologien: Eine Mediengeschichte der Schwarmforschung* (Zürich: diaphanes, 2012), 176. [My translation. All translations mine].

63 Lovecraft, 'The Whisperer in the Darkness,' 229.
64 Lovecraft, 'The Whisperer in the Darkness,' 257.
65 Hortense J. Spillers, 'Mama's Baby, Papa's Maybe: An American Grammar Book,' *Diacritics*, Vol. 17, No. 2. Culture and Counter-Memory: The 'American' Connection (Summer 1987), 67.
66 Lovecraft, 'The Whisperer in the Darkness,' 262.
67 Lovecraft, 'The Whisperer in the Darkness,' 250.
68 Serres, *Genesis*, 7.
69 Seb Franklin, 'The Context of Forms,' *World Picture*, Vol. 11, Summer 2016, 2.
70 Franklin, 'The Context of Forms,' 12.
71 Lovecraft, 'The Whisperer in the Darkness,' 221.
72 Lovecraft, 'The Whisperer in the Darkness,' 223, 228 & 230.
73 Lovecraft, 'The Whisperer in the Darkness,' 231.
74 Nancy, *Listening*, 14.
75 Jacques Attali, *Noise: The Political Economy of Music*, trans. Brian Massumi (Minneapolis & London: University of Minnesota Press, 2006), 6 & Serres, *Genesis*, 13.
76 Michel Chion, *Sound: An Acoulogical Treatise*, trans. James A. Steintrager (Durham & London: Duke University Press, 2016), 15.
77 Serres, *Genesis*, 15.
78 Serres, *Genesis*, 7.
79 Lovecraft, 'The Whisperer in the Darkness,' 234–235.
80 Friedrich Kittler, *Gramophone, Film, Typewriter*, trans. Geoffrey Winthrop-Young & Michael Wutz (Stanford: Stanford University Press, 1999), 3.
81 Kittler, *Gramophone, Film, Typewriter*, 86.
82 Attali, *Noise*, 92.
83 Lovecraft, 'The Whisperer in the Darkness,' 228.
84 Theodor Adorno, 'Punctuation Marks,' trans. Shierry Weber Nicholson. Accessible via: http://ubu.com/papers/Adorno-Theodor-W-Punctuation-Marks.pdf [Accessed 3 May 2022], 300 & 301.
85 Adorno, 'Punctuation Marks,' 300.
86 Adorno, 'Punctuation Marks,' 300.
87 Lovecraft, 'The Whisperer in the Darkness,' 233.
88 Adorno, 'Punctuation Marks,' 301 & 304.
89 Adorno, 'Punctuation Marks,' 304 & 305.
90 Bhabha, *The Location of Culture*, 124.
91 Lovecraft, 'The Whisperer in the Darkness,' 234.
92 Lovecraft, 'The Whisperer in the Darkness,' 250 & 255.
93 Sylvia Wynter, 'Unsettling the Coloniality of Being/Power/Truth/Freedom: Towards the Human, after Man, Its Overrepresentation—An Argument,' *CR: The Centennial Review*, Vol. 3, No. 3 (Fall 2003), 260.
94 Lovecraft, 'The Whisperer in the Darkness,' 263 & 259.

4 Hotel-Daddy-Wasp-Machine

The focus of this chapter is *The Shining*, both Stephen King's novel (1977) and Stanley Kubrick's 1980 film adaptation, treated as if they formed an 'interbeing,'[1] the one an outgrowth of the other. The argument is as follows: the affective relationship between Jack Torrance, played by Jack Nicholson in the film, and the Overlook Hotel, where he is the winter caretaker, yields a becoming-insect on the part of Torrance, entranced: the Hotel, mesmerizing, is coded as insectile. Once there with his family, depending on the job to see them through financial hardship, Torrance—alcoholic, bad-tempered, inadequate, entitled—enters into an alliance with the Overlook's 'dark nest,'[2] gradually yielding the assemblage of this article's title: Hotel-Daddy-Wasp-Machine. In *The Shining*, becoming, unlike its non-hierarchical articulation in Deleuze and Guattari's *A Thousand Plateaus* (1980), functions as a realignment with power. At the Overlook, Torrance receives commands concordant with the discourse of an 'honoured' place and 'just' retribution, in terms of either rewards and recognition (for him) or punishment (for anyone wanting his balls[3] or trespassing against his will). In Deleuzo–Guattarian vocabulary, which I assemble here in a sort of glossary of recurring terms: rather than deconstructive of the 'majoritarian' subject—the subject of power, a 'molar' entity—becoming is here patriarchised and accrues 'consistency,' leading back to form after all.[4] This consistency is distributed like swarms of wasps—the Overlook is wasp-force—not usually considered to be 'oedipalised'[5] creatures, recruited to the cause of phallic subjectivity. The event of becoming, rendered through 'demonic' wasps but executed in the name of the Father, is 'overcoded' with form, its elements 'reterritorialised' or 'facialised' to indicate precisely the restoration of form or figure.

I propose the interbeing (novel, film) in order to 'shine' on the processes of becoming developed in *A Thousand Plateaus*, which in turn allows us to think through the occurrences in *The Shining*. The reading I perform situates *The Shining*—perhaps a little illegitimately or parasitically—in the chapter on becoming in *A Thousand Plateaus*, whose subsections are variously identified as 'memories,' acquired, first, by particular subjects (a moviegoer, a naturalist, a Bergsonian, a sorcerer, etc.) and

DOI: 10.4324/9781003322757-4

subsequently emanate from movements, haecceities or molecules, points or blocks, even secrets, whose contents are too big for their forms.[6] I suggest a missing or hidden supplement to this particular chapter in *A Thousand Plateaus*—memories of a Hotel-Daddy-Wasp-Machine—concentrating on providing an interpretation, or revealing a blind spot, of becoming that establishes form instead of undoing it. Becoming should not automatically, as it were, be assumed to be anti-form and anti-fascist (frequently the dominant, seductive discourse), but considered in its transformations all the while staying alert to the danger of its being swept up in forms, that is, rigidly organised subject formations and a politics of the 'proper'. I understand the latter—at work throughout *The Shining* as well as in the contemporary world—as a politics of 'rightful' belonging, assumed to be 'owed' or 'deserved' by those whose subject status is never in question but who, yearning for a hierarchy and control denied or lost, feel conspired against, held back, dragged down. *The Shining* is a ghost story whose source of terror is a white daddy death cult: white men demanding, and pounding into being, a regime established on their eternal sovereignty. The object of this brutalising violence, both physical and psychic, are the preterite—women, children, racialised subjects—'fucking up'[7] the phallic order. What we witness in *The Shining*, where a perverse terminology of restitution endlessly circulates, is what happens when becoming infers or resurrects form, specifically the form of hegemonic, by default white, masculinity, 'rightfully' rendered and operating through the currents of a becoming-insect.

This chapter is divided into three sections, beginning with an investigation of Deleuze and Guattari's concept of becoming mapped onto *The Shining* and expounding the political motivations for the argument. This part ('Forms' of Becoming) also already gestures towards the phenomenon of fascination, manifested in the face and central to the subsequent section ('Dark Space, Play Face'), while the final section ('Sound Souls') concentrates on the Overlook Hotel's gaze which, as Lacan reminds us, does not necessarily involve the organ of sight.[8] The Overlook is legion, a composite of patriarchal and capitalist/settler colonial violence apparent in novel and film. The former establishes the patrilineage of abuse—which itself exceeds *The Shining* and reverberates throughout King's work as a whole—while the latter situates the Hotel, built between 1907 and 1909, on top of an 'Indian' burial ground, participating in and perpetuating the genocidal violence the settlers used to found their nation. The Colorado Lounge, the Hotel lobby with its Navajo and Apache designs—native American artefacts are on display in other parts of the Overlook, too—'rudely [incorporates],' as Roger Luckhurst observes, the culture it has exterminated. Luckhurst continues to argue that we are invited to read the Lounge, and the Hotel in general, historically though the tracking shot emphasising the vastness of this interior space, evoking the buried atrocities that constitute the United States and national white manhood.[9]

In the novel, the metaphor that organises the narrative, including the repressed history, of the Overlook is a wasps' nest, heard, in the film, through the music: 'secret wasps'[10] conceptualise an invisible force that enters Jack Torrance's psychic archive. The reading below depends very much on the interstitial (the overlooked and underplayed), a defining feature of becoming, something that happens between relations, disrupted selves. The schizoanalytic approach to *The Shining*—the methodology which I adopt—amounts to making a rhizome, integral to this essay's prose, a type of nesting, too: the wasps discovered, everywhere, on revisiting *The Shining*, on the pages of the novel, in the musical content of the score music. In Kubrick's film, swarms of wasps suggest themselves through the works of Béla Bartók, György Ligeti, Krzysztof Penderecki, Wendy Carlos and Rachel Elkind. Rather than 'outside' of the narrative, in a non-diegetic space, these sounds are instead acousmatic, which Michel Chion, in *The Voice in Cinema* (1999), defines as 'a sound that is heard without its cause or source being seen,'[11] without being able to be tied to a particular subject and, specifically, face. This source in *The Shining* is unseen, because faceless, occupying the entire field of vision: the Overlook Hotel is acousmachine, a 'voice-being' that comes from everywhere and 'is impossible to defuse'.[12]

My concern throughout this chapter lies in tracing the insectile, pertaining to an entomological fascination, across novel and film. If the first section is designed as a guide navigating through and against established ways of reading Deleuze and Guattari, scandalously paired with Lacan, the subsequent section is alert to the ways in which becoming-insect functions in relation to Torrance's absorption into Hotel-space. This interpretation hinges on a particular understanding of, and reaction to, spatiality developed by Roger Caillois. In 'Mimicry and Legendary Psychasthenia,' Caillois speaks of the reciprocal topographical organisation between subject and environment observed in the behaviour of mimetic insects, whose mechanism of assimilation to space mirrors the dispossession of the 'legendary' psychasthenic subject, unable to distinguish itself from its surroundings. In his article, Caillois notes that the 'epiphenomenon' of becoming-space is compelled by fascination,[13] an observation deployed in relation to Jack Torrance, fascinated by Hotel-space and adapting to the spatiality he finds himself caught in. Fascination is, of course, read in the face, just like subject formation occurs by way of the face. Who gets to have a face is equivalent to asking who gets to be a subject, recognised because of 'specific faciality traits'[14] that make up the signifying system of the face. Deleuze and Guattari describe the operations of the 'faciality machine'—really a 'White Man' machine; it recognises only the 'man-standard'—as overcoding apparatus.[15]

When a body is 'facialised,' it is assigned value as a subject. Processes of overcoding are, as such, linked to the face, to attempts at making the face and by extension the subject legible in the eyes of the Law/State.

Becoming-insect in *The Shining* is similarly legible in the face: Torrance's fascination, freezing his face into a smile, both corresponds to and further generates a becoming-insect, a metamorphosis stimulated by his mode of engagement with Hotel-space. This particular configuration, occasioned by Caillois's article, understands Torrance as site where the subject, like the mimetic insect, organises itself topologically according to the Overlook's weird coordinates. Becoming-space, the distribution of subjectivity in relation to the 'space-agency'[16] of the Hotel, is hence an entomological phenomenon also describing the affective becoming of the 'human' (Torrance, in this case). As a result, becoming-space = becoming-insect, though the latter further manifests itself through a phagocytosis, as it were, of the subject,[17] whose assimilation is captured, in King's novel, in an apparent facelessness, emergent 'behind' false faces or the impassively faced. Midnight, as King writes, is the time to unmask, when 'everybody exposed faces that were those of rotting insects.'[18] Although easily interpreted as an anthropocentric and racist event—the 'White-Man face' assaulted by the total other—in *The Shining* becoming-insect functions as process for the recalibration of an implacable master subjectivity. Rather than posing a threat to the masculine order, becoming-insect materialises an absolute, diffusive white sovereignty, a mode of an overwhelming phallic and necropolitical organisation picked up through *prise de son*, repurposed to study the voice, usually 'there to be forgotten about in its materiality'.[19] In cinema, *prise de son* tends to be used so as to drown out everything that interferes with the 'human' voice, privileged over other sounds. The last stage of this analysis, however, foregrounds the swarming 'sound souls'[20] of the Hotel, whose material presence and wasp-voice refer back to the 'macroface,'[21] everywhere and omnipotent, of the Father.

'Forms' of Becoming

In *A Thousand Plateaus*, becoming functions as an un-doing of the subject without regression or, conversely, progression. Becoming stages an 'involution' as opposed to an evolution[22] and can never be captured by the verb 'is' succeeded by an imago, to borrow an entomological term, the final stage in an insect's metamorphosis. It has 'no term,'[23] by which point something has been reached or settled, but concerns alliances effectively ending the form of the subject: subject becomes event. A brief account of becoming according to Deleuze and Guattari includes an 'irresistible deterriotrialisation' that affects the subject, throwing it out of its Oedipal drama; the expansion, as it were, of the thing called the 'human' into the more-than-human; pack affects or 'unnatural' assemblages replacing 'family feelings or State intelligibilities,' making possible a politics of 'minoritarian groups,' that is, the preterite, subaltern, those in revolt, who are extrinsic.[24] The process dissolves form or the organism of the coherent subject by way of 'becoming' relations between organs, functions, particles. These enter into a 'zone of proximity'[25] or a haecceity, a composition of affective intensities that disrupt the molar or majoritarian entity, a subjectivity defined by form, which is rooted and closed.

In a section on becoming-music and the refrain, Deleuze and Guattari acknowledge '[t]he fact that there is no deterritorialisation without a specific reterritorialisation.' This realisation should subsequently 'prompt us to rethink the abiding correlation between the molar and the molecular,' between that which is rigidly organised, unified, defined by form and that which is other, formless and unruly. Deleuze and Guattari continue: 'no flow, no becoming-molecular escapes from a molar formation without molar components accompanying it, forming passages or perceptible landmarks for their imperceptible processes'.[26] The molar always lies in wait, awaiting its irruption into the scene of becoming. Becoming or swarming, beyond Deleuze and Guattari and as expression of the body politic, has the potential to renegotiate the model of modern sovereignty, as Eugene Thacker argues in his two-part investigation on 'Networks, Swarms, Multitudes'.[27] Rather than located in a single place, person or thing, a swarming sovereignty, while dynamic, might still be predicated on some type of organisation, even if not centrally controlled, an articulation with which it nonetheless remains compatible. Lauren Wilcox's work on swarming and becoming-insect similarly shows that these organisational modes 'no longer only represent a line of flight outside of the masculinist politics of control' because 'appropriated for a necropolitics of surveillance and warfare'.[28] Figurations like becoming-insect can, then, come to behave as strategies of masculine sovereign power, even if these fly in the face of Deleuzo–Guattarian conceptions of becoming-other. The assemblage generated in *The Shining*, Hotel-Daddy-Wasp-Machine, operates in and according to such a regime, whose ultimate manifestation is the 'individuation' it produces. This individuation might be notated as molecular arrangement, with its predominance of hyphens, but is in fact molar unity.

At first glance, the story of *The Shining* is not material to be considered in conjunction to the 'minoritarian' politics of becoming. To begin with, there is no doubt that Jack Torrance experiences his becoming as anything but molar, as a 'becoming […] major'[29] realising his potential for dominance. In *The Shining*, becoming is phenomenon of crystallisation—literalised in Kubrick's film: daddy freezes to death—and is further brought about through a negative conception of desire incongruous with Deleuzo–Guattarian thinking, refuting the Lacanian concept of lack at the heart of the desiring subject. It is not, however, impossible to demonstrate a 'productive dialogue' between Deleuze and Lacan, as Boštjan Nedoh and Andreja Zevnik have shown, although that dialogue tends to circumvent the 'most irreconcilable moments'[30] such as the 'problem' of desire. I'm not proposing to resolve this problem; rather, I am putting it to work in order to show how becoming has been seized by masculinist discourses, so adept at appropriating tactics of resistance and creative energies, like the figuration of the swarm, to its cause.

The Shining lends itself to readings focused on the problematics of becoming even before entering its narrative realm, bearing in mind John Sears's argument in *Stephen King's Gothic* (2011). Sears re-reads King's immense, excessive oeuvre as textual machine whose identity (like that of genre more generally) always exceeds itself and is constantly reworked 'into new forms,'[31] though both the new and the idea of form are concepts that must be put under pressure. He understands King's writing as becoming, his novels as desiring machines, establishing Deleuze and Guattari as a compelling methodological framework with which to approach it. The 'logic' of becoming, bound up with practices of re-reading because of King's performance of genre, is also at work in *The Shining*, where it is simply articulated as such, a becoming contested in the face: 'faces are running, changing, becoming something pestilent'.[32] The face is zone of transformation, explicitly conceived as insectile, hosting 'rotting insects' or otherwise 'crawling' with 'heavy-bodied wasps.'[33] The process of becoming consequently happens at multiple levels or nodes, outside the text and inside it, while comprising the many proliferations continually supplementing the 'original' with additional dimensions: *The Shining* sprouts, returns, is parasitic, insists.

In his article 'The Unempty Wasps' Nest,' Graham Allen briefly considers the wasp, a 'sclerophthalmic animal [...] with hard, lidless eyes' and suggests that the Overlook Hotel offers up 'a mode of vision with sees everything and sees it all the time'. He notes that this vision is 'unbearable, because it is total, like the vision of a god, or like the vision of a movie-camera, another "being" that does not have eyelids to close, a sclerophthalmic machine if you will.'[34] Allen's focus lies elsewhere, in practices of reading, interpretation and adaptation open to chance, but his analysis of the Overlook's gaze, returning the trope of the wasps' nest in the novel, nonetheless identifies what makes *The Shining* so compelling. Indeed, it often yields an obsessive, paranoid interest, demonstrated, for example, in the 2012 documentary *Room 237*, featuring a number of more or less convincing and imaginatively argued approaches to the film. *The Shining* triggers a compulsion to repeat, urging acts of re-reading, viewing and interpretation, as if it constituted a traumatic neurosis. It is a text that multiples and persists in various forms, including a sequel to the novel (*Doctor Sleep*, 2013), also made into a film (2019); the 'reparative' TV mini-series (1997), scripted by King and therefore more faithful to the novel; an opera (2016); a wiki site; a *The Simpsons Treehouse of Horror* episode (1994); a sequence from Steven Spielberg's 2018 film *Ready Player One*; a 1993 single by the Dutch band *Hocus Pocus*; Jack Torrance's 'posthumously' published 'writing project,' *All Work and No Play Makes Jack a Dull Boy* (2008); a comparison by the German sports magazine *Kicker.de* which, prior to the so-called *Der Klassiker* in April 2022, renders Bayern Munich's then star striker Robert Lewandowski as Jack Torrance terrorizing Borussia Dortmund, captured as Wendy; Melania Trump's hair-raising White House Christmas decorations (2018 and 2019), compared, on Twitter, to scenes from the film: officially released perspectives of the Trump White House decorations are photoshopped to integrate the dead Grady girls or a close-up of Torrance's frozen face, not at all looking out of place in coldly symmetrical arrangements. A whole culture industry revolves around the Overlook Hotel as fascinating object: the Hotel is produced as such across these iterations. The Overlook is enigma which, like Medusa's head, suspends its onlookers in a gaze that interpellates by surprise, through sound, for example, displacing the subject into the other's field of vision.[35] The shining, or the shine, in fact partly functions as another word for fascination and is, at any rate, allied to its affective power, considering the fundamental passivity into which the shining/fascinated subject sinks. The encounter with the Overlook, fascinating object, is one which structurally calls for returns, not only because it remains enigma but because 'we' desire to be held captive in an eternal, sclerophthalmic, gaze.

My argument arises out of the dynamic between these things: the subject's (also mine/our) 'passion for the image'[36] of the Overlook Hotel, signalling integration into the Hotel's gaze and field of vision or desire; *The Shining*'s function as proliferating, parasitic text; because the planes of becoming in *A Thousand Plateaus* have to be supplemented with the memories of an assemblage left out of the picture. The constellation *The Shining/A Thousand Plateaus* provides a way to interject into a discourse on becoming that tends to assume its counteraction to State power—white supremacy and misogyny—and to take for granted its orientation towards an ethics of responsive engagement with the other. As Lori Brown has shown, the other body in the process of becoming is frequently abandoned by Deleuze and Guattari; she writes that the resources of care might be present in their work but that the commitment to the well-being of the other is missing.[37] Brown's purpose is to expand the ethical possibilities of their tenth plateau—the chapter on becoming—by 'gather[ing] methods' that enable such a commitment.[38] I follow suit by seeking the continual activation of the political and ethical potentiality of becoming, which has never lost its urgency. Alliances must be formed, but not on the basis of identity, identifiability or majoritarian positions, and all the while being alert to the mechanisms co-opting becoming to regimes of State, from which it is not, at any rate or forever, exempt.

Certain majoritarian transversals already exist in the tenth plateau, 'abject reterritorialisations,'[39] as Deleuze and Guattari put it, that haunt the dissolutions of form described and discovered across the different instances of 'remembrance'. Reterritorialising aspects occasionally enter passages of becoming in *A Thousand Plateaus*, as we have already seen.

I temporarily suspend some of the givens, the good, of becoming, namely that it cannot function as part of a molar structure, aggregating multiplicities so as to make them serve one Master. *The Shining* functions as a case study to think this suspension, the processes by which becoming becomes molarised. The proposition of becoming/swarming realised in this conjunction *A Thousand Plateaus* + *The Shining*, another interbeing, is that with which some of us are intimately familiar: the figuration (map of the living dead)[40] of total phallogocentric control. This unlikely alliance allows us to push further the potential radicalisation of political and ethical thought that Deleuze and Guattari's *A Thousand Plateaus* is turned toward but whose central concepts must nonetheless not escape scrutiny. It might well be that the suspicion of a non-sanctioned, somehow wilfully trespassing, intervention lingers, yet trespassing is itself so crucial to the demonic-paternal logic operating in *The Shining*. The politics of the 'proper' at the Hotel constantly invokes rules, often unwritten and obscure, regulations and contracts that must be observed by the Father-Caretaker on the lookout for those who are where they shouldn't be:

> Husbands and fathers did have a certain responsibility. Father Knows Best. They [women and children] did not understand. That in itself was no crime, but they were *wilfully* not understanding. He [Torrance] was not ordinarily a harsh man. But he did believe in punishment. And if his son and his wife had wilfully set themselves against his wishes *against the things he knew were best for them*, then didn't he have a certain duty—?[41]

My analysis, indexed, as it is, to an energy to trespass, braces itself against the law of the Father ('Father Knows Best') by paying attention to *The Shining*'s architecture or infrastructure of molarisation, that is, the Overlook Hotel, whose poetics is that of the swarm, usually perceived, not least in horror fiction, as destroyer of form. I am not, however, using *The Shining* to argue that horror fiction is generically 'predisposed' to save stable categories of identity—therefore imagining becoming as catastrophic—but to focus on an 'abiding' molarity repressed in the process of becoming.

Dark Space, Play Face

In *The Shining*, becoming-insect is initiated through a dynamic which can in a first instance be refracted through what Caillois terms the 'mutual organisation' or 'reciprocal topography' between mimetic insect and space, an event that 'digests' the insect into the world immediately around it.[42] He mentions the tactics of a type of butterfly, a Satyrid, flattening its wings so as to appear 'like a line almost without thickness, imperceptible, perpendicular to the flower where it has alighted.'[43] He subsequently links this force of assimilation to fascination and to personality disorders like schizophrenia, often experienced as disintegration (although Deleuze would call it connectivity). Caillois likens this phenomenon of the subject 'no longer [knowing] where to place itself' to a becoming-space, a state in which the apparent ontological distinction between self and other/ environment fails to hold. The question 'where are you?' addressed to a schizophrenic elicits an impression of being pursued by space to the point that it replaces subjectivity:

> Then the body separates itself from thought, the individual breaks the boundary of his [sic] skin and occupies the other side of his senses. He tries to look at himself from any point whatever in space. He feels himself becoming space, dark space where things cannot be put.[44]

Caillois is interested in the curious quality of relationship between subject and space. His approach proceeds from the perspective of the organism or self concerned, not from a space that somehow exists—over there—prompting mimesis or assimilation. A butterfly reacting to a flower produces the assemblage Satyrid-flower, to borrow Deleuze and Guattari's notation, to the point that it becomes impossible to assign attributes to the one and not the other. To clarify: it is not as if the subject, without desire, were at the mercy of a fascinating object or space that on its own—without the interference of the on-looking entity at the other end—exerted its 'rays'. Rather, there is an interplay between subject and object already caught in desiring machines sweeping along the integrity of the I. If the I responds to the fascinating object/space by growing rigid and being rooted to the spot—*viz* Freud on Medusa[45]—then that mechanism of defence is not simply a negation of fascination. The ego cannot cancel out the fascinating element because fascination is or collapses into desire. As such, it constitutes a movement that acknowledges a relation between subject and object. In other words, fascination is not that which belongs to the object, whose 'oral eye'[46] wants to engulf me, but emerges over there, because over here I seek its recognition, just as I recognise myself in the object. The subject–object assemblage, like the Satryrid-flower, forms an aggregation of desire or, in Caillois's vocabulary (which is also the language of psychoanalysis), the I becomes 'convulsive possession'[47] of the fascinating object/space.

To be 'convulsively possessed' by space persists in genre fiction and particularly in King's work, often relying on and amplifying the 'magical hold'[48] of locations which 'destructure' and redirect the subject's desires.[49] Although such sites, like the Overlook Hotel, do not exert the same fascination on every subject, their power is nonetheless considerable, further activated by the subject it can most easily claim. Even Kubrick's film, on the face of it denying the 'bad place' motif of its source material, engages with it, if less evidently so. The smile floating between novel and film codes Torrance's 'turn' into 'dark space,' despite existing, in the film, right from the start. The close-up of Jack's face, used repeatedly once ensconced at the Hotel, renders the smile as rictus, the expression *par excellence* of his fascination. In a still from the film, Torrance's face hovers over the miniature replica of the Overlook's hedge labyrinth outside, a curious smile etched into the corners of his mouth, his gaze riveted to an elsewhere. Following Caillois, fascination is the phenomenon that assimilates insect and (schizophrenic) subject to space. More generally, fascination, according to Freud in his study on Leonardo da Vinci and Lacan in *Seminar II* (1954–1955), is fundamental to subject formation, often producing an excess of identification because fascination depends on the capacity to return it, to mimic it. Freud focuses on the lost smile of the mother, compelling Da Vinci to transfer it to every face he paints: it is indicative of fascination, an expression that, in this iteration—'[cold] and [lifeless]'[50]—means being transfixed, called away from the world. The smile is *Brennpunkt*, to borrow Esther Leslie's description of a 'splinter of space and time,'[51] that pivotal moment which in a photograph draws and arrests the eye. Similarly, it is Torrance's smile in one of the most predominant stills from the film that registers his capture by the Overlook Hotel.

Unlike Freud's reading of the Gioconda, though, Torrance's smile infers his father, fleshed out in the book (in its original release, the film only hints at a history of paternal abuse). Daddy is a difficult figure to identify with because his violence—the terrible domestic assaults—is initially unpalatable for 'Jacky-boy,' though he (like Danny, Jack's telepathic son, caught in the same Oedipal drama) loves daddy, Janus-faced, too. It takes the Hotel to fully awaken the '[slumbering]'[52] soul within, so that becoming-space brings about the total identification of 'Jacky' and daddy, arriving at the embodied resurrection of the despot-machine. In his *Seminar* on the ego in Freud's theory, Lacan argues that the subject only 'gains its unity in so far as it is fascinated'.[53] That unity, in *The Shining*, is the Overlook Hotel, organising Torrance's subjectification as a reflection of its order. Becoming-Hotel, cutting a smile into Jack's face, promises to restore subject unity, to rescue it from failure: the failure of a wannabe writer (Kubrick's film) or writer (King's book), of a husband and father to 'live up to his responsibilities' and discipline 'wilful' wife and child. Gouged exclusively into the face of the father, not just Torrance's but Grady's before him, the smile is associated with 'correction,' the expression proper to the phallic father:

> 'He [Danny] needs to be corrected, if you don't mind me saying so. He needs a good talking-to and perhaps a bit more. My own girls, sir, didn't care for the Overlook at first. One of them actually stole a packet of matches and tried to burn it down. I corrected them. I corrected

> them most harshly. And when my wife tried to stop me from doing my duty, I corrected her.' He [Grady] offered Jack a bland, meaningless smile. 'I find it a sad but true fact that women rarely understand a father's responsibility to his children. Husbands and fathers do have certain responsibilities, don't they, sir?'[54]

The bland smile is expression or impression of fatherly 'duty,' the shine of his murderous corrections. If the Satyrid in its perpendicular assemblage with the flower offers only a 'minimum surface'[55] to an observer to be seen in an otherwise imperceptible event, the smile in *The Shining* constitutes that surface, the lines of the mouth the site of a visible approximation between subject and object. Becoming-space or becoming-Hotel, encompassing the movement Deleuze and Guattari did not think possible, that is, becoming-daddy, forms part of a complex manoeuvre that figures as a becoming-insect totally shaped by the law of the Father. The process of becoming, whatever its molecular possibilities in different circumstances, has thus far arrived at the formation of a Hotel-Daddy-Machine, an assemblage integrating Chion's concept of the acousmachine, inhabiting all points, 'forever and ever,' in space and time. From here on in, I am going to concentrate on the 'subsequent' stage of becoming taking place in *The Shining*, though the impression of a sequence of stages is misleading: the aggregation of the Hotel-Daddy-Wasp-Machine really is launched from the start. Neither one of these 'intensities' circulates without the other.

Sound Souls

The site of convergence of the 'mutual organisation' between Torrance and Hotel is the face, forming playing field with the wasp-force of the Hotel. While the preceding part of the chapter paid attention to the smile as indicator of the fascinated, overcoded subject, this part is concerned with facelessness, predominant in King's imagination. I take facelessness to mean and function in the service of the 'macroface,' the face of the Father, whose 'centre is everywhere and circumference nowhere'.[56] This macrofaciality—phallogocentric power machine—is produced and organised by the overcoding Hotel, inscribed as insectile. The section, consequently, is attuned to how sound realises the swarm-souls of the Overlook, a representation that is evident, almost unnoticeably, throughout King's novel, introducing the metaphor of the wasp as invading/occupying force way before the discovery of the nest in Part III. Gathered from the book's interstitial areas, a pack mode of insectility emerges in processes of (compulsive) re-readings:

> advice that stings, 28; nesting, 34; slow-moving, mean insects, 103; nest as workable symbol for Jack Torrance in passive mode, 104; hand impaled by needles, 106; insectile forms rise into the air,

> droning, 126; nest like deadly fruit, 131; secret wasps, 164; the sudden, needling sting, 173; moving wasp, having stung 179; the wasps [...] the play, 197; the thought came from outside, insectile, buzzing, softly cajoling, 202; what dark nest, 228; hurt-think like wasps at night in my room, 234; dry pile of papers like wasps' nest, 255; monstrous mechanized wasp, 258; the shout to unmask, everybody exposed faces that were those of rotting insects, 300; hear the ghosts: the somnolent hum of summer wasps in a ground nest, sleepy, deadly, beginning to wake up/the sound souls of the hotel, 306; buzz of wasps & wires, 310; this is what's like, to stick your whole hand into the nest, 333; know where the wasps are, 354; high, insectile buzzing sound, 383, 387; heavy-bodied wasps crawling over decaying face, 392; the insectile sound of the motor, 395; the blasted nest: and, looking back over his shoulder, as he was now, he had on that day seen a dark cloud of hornets rising into the hot air, swirling together, breaking apart, 407; the single group intelligence could sting to death, 408.

Becoming correlates, as Deleuze and Guattari write, to that which swarms and stands in opposition to the 'man-standard.' In *The Shining*, however, 'man'—an awful incarnation without limit—is produced by wasps operating as molar system, the Hotel's politics of masculine control. It is, on this note, significant that in the book, prior to his stay at the Hotel, Torrance had published a few short stories, the form that, according to Frank O'Connor, is intensely aware of 'human loneliness'.[57] King's Torrance, having lost his teaching job for beating up a student, with a similar record of abuse at home, is—exclusively because we are privy to his haunted interior life—a figure who would write in a form reflective of his own life of ghosts. Once at the Overlook, though, he dreams of compiling a long, explosive book on the history, in its entirety, of post-war America, situating the Hotel, the key to all mythologies, at its dead centre: a perspective overlooking, in command of, the field of representation, no longer shambling along on its margins.

Swarming has here undergone a 'material transformation,'[58] stinging the 'man-standard' into place. In-human mechanism for the desire—demonic paternal authority—of the Hotel, Torrance functions as storage apparatus whose ancestral information is retrieved and which, smiling blankly, cites the laws of the Overlook Hotel. The Overlook's 'wasp-like buzz' activates or disturbs the dark nest of the Father. Consequently, the 'outside'—despite Danny's perception of 'hurt-think'[59] as exterior, wasp-like thought—is internalised, archived as always already internal. The overcoding that the Hotel effectuates refers to the erasure of any kind of inscription that does not conform to its codes nor is designed to carry out its orders, rendered as the 'somnolent buzz'[60] integral to Torrance, discovering his unitary mirror-machine in the Overlook Hotel.

The wasps' nest as visual trope is missing in Kubrick's film, but the 'deadly siren song'[61] of the Hotel is absent presence. The Overlook is insectile voice-being, the score music indicative of the 'speech' of an acousmachine, deriving from Chion's analysis of the 'acousmêtre':

> When the acousmatic presence is a voice, and especially when this voice has not yet been visualized—that is, when we cannot yet connect it to a face—we get a special being, a kind of talking and acting shadow to which we attach the name *acousmêtre*.[62]

The acousmachine, like HAL in *2001: A Space Odyssey* (1968), is voice without body,[63] without face. Coming from everywhere, the acousmachine inhabits everything and is in this respect affiliated to Deleuze and Guattari's concept of the macroface. The Hotel's voice is sound 'without' semantic content or, at any rate, existing outside of language or information signals 'we' understand. Yet, as Chion proposes in *The Voice in Cinema*, 'there is no soundtrack' (*il n'y a pas de bande-son*),[64] a statement that echoes Derrida's '*il n'y a pas de hors-texte*,' because there is no image that is not affected and animated by sound, even if sound is often swallowed up by the image. To dispense with 'the hierarchy of perception,'[65] subordinating all sounds to the image as well as to the 'human' voice and drowning out other remains that form the sonic terrain of a film, privileges that which is or becomes deafening in *The Shining*, that is, the acousmachine of the Overlook. Without face, the acousmachine—itself 'outside the image, and at the same time *in* the image,'[66]—is macroface, whose 'voice' is the place, the dimensions of the Hotel itself. There is nowhere to go, nowhere to be, where the Overlook is not.

In his biography of Stanley Kubrick, Vincent LoBrutto reports Wendy Carlos's reflections about the score music in *The Shining*: 'Stanley wanted these sounds that would sneak in and go by. He called them low fly-bys. They were sounds that would sneak into you, subconsciously'.[67] These low fly-by sounds operate exclusively on the level of the score music, the swarm sounds and heartbeat of the Hotel acousmachine. The source music, on the other hand, diegetic music that belongs to the film's *zone visualisée*, briefly construes a precarious safe zone: William Lava's *Painted Sea Storm*, for example, plays on the television showing a cartoon in the opening scenes. Stephan Sperl notices how 'untroubled [*sorglose*] compositions'[68] like *Masquerade* by Jack Hylton and his Orchestra and Ray Noble's *Midnight with the Stars and You* refer to the Overlook revving into 'life'—the tune returns in the final scene, moving in on Torrance's smile arrested in the photograph from 1921—while *It's All Forgotten Now* accompanies Grady in the blood-red bathroom, failing to remember murdering his wife and kids. The Overlook overcodes the messages of seemingly innocuous music, enfolding it into its infernal, core-less system. There is no dimension left untouched by the Hotel, whose voice is 'king sound,'[69] sneaking inside from a position that appears *hors-champ* to one that has always occupied everything in sight.

The entire sound field is subordinated to this sovereign, but by no means isolated, sound. A dense, stratified, 'low and slowly moving'[70] cluster of notes corresponds to King's description of the 'somnolent hum of summer wasps in a ground nest, sleepy, deadly, beginning to wake up,' which, Sperl suggests, mirrors the labyrinthine architecture of the Hotel.[71] The score music, the acousmatic voice of the Overlook, is the sound, an 'aural wash,'[72] that sets the tone and undoes time. Sound sequences, to once again gesture to Chion's work, 'often provide what might be called a certain temporal tonality,'[73] a way of traversing time, of feeling it unfold. This duration at the Overlook is perverted 'shimmering time,' usually a pleasant temporal tonality which roots to the spot and is spellbinding, capturing the spectator in filmic events.[74] To be frozen, fascinated, assimilated: that's shimmering time at the Overlook, where time either stands still or, like the sounds that give it tonality, is all mixed up: Tuesday, Saturday, 4 pm, 1945, 1921. The voice draws 'you' in, held spellbound in shining time, by low fly-by sounds that imperceptibly produce a field of intensities, a vibrational, corridic infrastructure as politics of capture.

The age of insects (a misnomer, really, for the reign of the Father) is inaugurated through sound in the film, which yields a becoming traced back to a centre of unification. The works by Bartók, Carlos and Elkins, Penderecki and Ligeti forge a climate of becoming under the watchful, unanchored eye of a voice-being that, though distributed and extending into every available space, remains an absolutely sovereign agency. 'I sometimes wonder,' writes Luckhurst, 'if this buzzing' generated by Ligeti's *Lontano* 'is the sound of the wasps' nest [...] meant to evoke the hive mind of the Overlook Hotel, agitated with inhuman intelligence,' continuing that Penderecki's *Polymorphia* 'sounds like a horde of insects eating their way out of the string section'.[75] The score music, used to designate shining time, that arresting, spellbinding and insectile tonal temporality, orients Torrance towards 'becoming [...] major.' To all these movements, there is a disturbing, asphyxiating closeness, even though they function as the motif for telepathy—communication, after all, at a distance. *Lontano*, purely by virtue of its name, is case in point. Italian for far away, *Lontano*, like those other aural phenomena, is sound as heartbeat, the manifestation of a vitally unliving sensibility, sentient yet undead and eternal, at once attractive and repellent, distant and claustrophobically near. Pulsations—the impression of sounds like wasps flying by—close in on their target, disturb it, tone it to the Hotel's singularity of purpose: this force is in command of the whole texture of the film.

Torrance, entranced, 'enters into composition'[76] with the Hotel. This movement is a glide, continuous, automatic, like Danny's space exploration by tricycle, itself followed by the smooth machinic gaze of something that is not an entity, as Garret Brown characterised his Steadicam.[77] The 'territory sounds' of the Overlook, characterising its locale, are swelling and abating, simultaneously ambient—filling the space of the film while spilling beyond it, into the theatre where 'we' sit, immobilised—and internal: these sounds are unbound by space.[78] They have also always been there, like Grady, Torrance, the law of the Father, making it impossible to express or experience duration. Swarms of sounds, music of trance and terror, un-form the linearity of time, which is cyclical, like a refrain, spinning on an axis that returns the same acts of violence: 'in the Overlook things just went on and on'.[79] Ligeti notes that *Lontano* functions as

> a piece of music which gives the impression to continually flow along [*dahinströmen*], as if it had no beginning, and no end; what we hear is essentially only a section of something that has always already begun [*das schon immer angefangen hat*] and that will continue to reverberate forever.[80]

The enigmatic sounds of the Hotel are indicative of its forever; always present, they reterritorialize the fascinated subject as apparently faceless. In King's work, as Sears notes, horror is 'unfaced,'[81] a dissolution of form that, like the secret wasps, can be glimpsed throughout *The Shining*:

> a cabinet, lying on its face, 35; Wendy dreams of her mother's face, 51; haunted by faces, 52; Danny born with no face, 54; unremembered faces, 56; the rabbit's face which, close up, looks like no face at all, 192; cane smashing against face, 212; Jack's true, hidden face, one of despair, 216; faces carefully set, 218; misted faces, 223; zombie-like face of a stranger, catatonic, 227; there was not a face, only a mask of blood, 256; a fright-mask, 269; blank face, 272; as-yet-unseen face, 287; faces like those of rotting insects, 300; faces that are running, changing, becoming, 322; what famous faces, hidden behind masks, 331; deface the furnishings, 388; it wore many masks, but it was all one; it was hiding behind Daddy's face, 391; false face, 392; ripped face, 397; what remained of the face became a strange, shifting composite, many faces mixed imperfectly into one, 400.

Sears argues that the faceless other is situated as feminine in King's writings, largely at hand of *Pet Sematary* (1983) and *It* (1986), invariably marking the gender-neutral pronoun 'it' as a 'she.' *The Shining* does not figure as part of the selection of texts considered in his chapter, because the novel does not fit with an interpretation that is, no doubt, indicative of King's work as a whole. The faceless entity in *The Shining* is the *Ungestalt* of the 'man-standard,' a disassembly that precedes a more terrifying remembering (working in both senses of the word) of the Father. The becoming-insect taking place in *The Shining* is absolutely bound to the despotic agency of the Father, materialising as insect-thing or, more precisely, as Hotel-Daddy-Wasp-Machine, 'restructuring' Torrance's face in its image.

By way of a conclusion, I'd like to turn or return to the following tableau: on 2 June 2020, the day after his visit to St. John's Episcopal Church, where he had protestors, including the clergy, teargassed for taking to the streets against systematic racism and the murder of George Floyd by Minneapolis police, Donald Trump posed for a photo at St. John Paul II National Shrine in Washington, DC. Donald and Melania Trump stand facing a crucifix, frowning into the camera, arms rigidly at their sides. In the background of an image-edited picture circulated on Twitter,[82] the dead Grady sisters lurk, framed in the blue-curtained entrance hall reminiscent of the arresting, oneiric perspective of the Overlook's blood-gushing elevator doors. Eerie blue light partly shines through the fabric of the drawn curtains, cascading down the left of the image: echoes of the Overlook Hotel. As already mentioned, the Trump White House ('Here's Donny!') keeps calling up references to *The Shining*, staging the perpetual recurrence of the same thing, that is, the horror of the molar entity, enfleshed, here, by the former President of the United States. The edited picture testifies to *The Shining*'s enduring deployment as source to be cited in the contexts of necropolitical control, commenting on the death-grip of white patriarchal power structures and, at the same time, drawing attention to its behaviour as perpetually parasitic text.

As Michel Serres argues, a parasite or signal—he also calls it noise, the sound souls, if you wish, of another order—entering a system does not necessarily mean the system's cessation or alteration. A parasite is a 'function of time,'[83] able to cause only a momentary interruption, after which the system resumes its smooth flow. The system (Trump's presidency) does not break down as a result of this particular intrusion (the Grady girls) but is confirmed in its demonic power. The uninvited guests, who (don't) belong, are, however, also capable of switching the relations between host (the original photo) and interrupting parasite (*The Shining*). The ghost-girls at the back of the picture are symmetrical, doubling each other as much as the couple in front, whose posture—arms arranged along their bodies—they mirror, suggesting that it might well be the former President and his wife that are the terrible apparitions to be transplanted as ghouls into other scenes, the corridors of the Overlook Hotel. The relations between parasite and host are revealed to be interchangeable in this instance: host and parasite are '*simpático*'.[84] In different circumstances, though, with respect to the argument laid out above, *The Shining* retains its parasitical signal, prying open a passage between the molecular and an abiding molarity in the dynamics of becoming. Remembering, also, the hold that *The Shining* has over its addressees, sinking into the Overlook's gaze, the insectile as operational mode of transformation at the Hotel might, after all, parasitically affect the patterns of becoming as they are described in *A Thousand Plateaus*.

Notes

1 Gilles Deleuze & Félix Guattari, *A Thousand Plateaus: Capitalism and Schizophrenia*, trans. Brian Massumi (London & New York: Continuum, 2004), 27.
2 Stephen King, *The Shining* (London: Hodder & Stoughton, 1990), 228.
3 King, *The Shining*, 178.
4 These terms are all used by Deleuze & Guattari in *A Thousand Plateaus* to indicate the subject of power. See pp. 30 & 263.
5 Deleuze & Guattari, *A Thousand Plateaus*, 257.
6 Deleuze & Guattari, *A Thousand Plateaus*, 316.
7 Stanley Kubrick (dir.), *The Shining* (Warner Bros., 1980).
8 Jacques Lacan, *Four Fundamental Concepts of Psychoanalysis*, trans. Alan Sheridan (New York & London: W.W. Norton & Co., 1998), 84.
9 Roger Luckhurst, *The Shining* (Basingstoke & New York: Palgrave Macmillan, 2013), 43 & 44.
10 King, *The Shining*, 164.
11 Michel Chion, *The Voice in Cinema*, trans. Claudia Gorbman (New York: Columbia University Press, 1999), 18.
12 Chion, *The Voice in Cinema*, 25 & 43.
13 Roger Caillois, 'Mimicry and Legendary Psychasthenia,' trans. John Shepley, *October*, Vol. 31, 1984, 25.
14 Deleuze & Guattari, *A Thousand Plateaus*, 186.
15 Deleuze & Guattari, *A Thousand Plateaus*, 197 & 199.
16 Jussi Parikka, *Insect Media: An Archaeology of Animals and Technology* (Minneapolis & London: University of Minnesota Press, 2010), 101.
17 Caillois, 'Mimicry and Legendary Psychasthenia,' 30.
18 King, *The Shining*, 300.
19 Chion, *The Voice in Cinema*, 1.
20 King, *The Shining*, 306.
21 Deleuze & Guattari, *A Thousand Plateaus*, 197 & 233.
22 Deleuze & Guattari, *A Thousand Plateaus*, 263.
23 Deleuze & Guattari, *A Thousand Plateaus*, 262.
24 Deleuze & Guattari, *A Thousand Plateaus*, 257, 262, 266, 271 & 272.
25 Deleuze & Guattari, *A Thousand Plateaus*, 301.
26 Deleuze & Guattari, *A Thousand Plateaus*, 334.
27 Eugene Thacker, 'Networks, Swarms, Multitudes,' Parts One and Two, *C-Theory*. Available via: http://ctheory.net/ctheory_wp/networks-swarms-multitudes-part-one/?template=print; http://ctheory.net/ctheory_wp/networks-swarms-multitudes-part-two/ [Accessed 10 October 2019].
28 Lauren Wilcox, 'Drones, Swarms, and Becoming-Insect: Feminist Utopias and Posthuman Politics,' *Feminist Review*, Vol. 116, No. 1 (2017), 28.
29 King, *The Shining*, 251.
30 Boštjan Nedoh & Andreja Zevnik, 'Introduction: On a Disjunctive Synthesis between Lacan and Deleuze,' in *Lacan and Deleuze: A Disjunctive Synthesis*, ed. Nadoh & Zevnik (Edinburgh: Edinburgh University Press, 2017), 1.
31 John Sears, *Stephen King's Gothic* (Cardiff: University of Wales Press, 2011), 94.
32 King, *The Shining*, 322.
33 King, *The Shining*, 300 & 392.
34 Graham Allen, 'The Unempty Wasps' Nest: Kubrick's *The Shining*, Adaptation, Chance, Interpretation,' *Adaptation*, Vol. 8, No. 3 (March 2015), 366.
35 Lacan, *Four Fundamental Concepts of Psychoanalysis*, 84.
36 Maurice Blanchot, *The Space of Literature*, trans. Ann Smock (Lincoln & London: Nebraska University Press, 1989), 32.

37 Lori Brown, 'Becoming-Animal in the Flesh: Expanding the Ethical Reach of Deleuze and Guattari's Tenth Plateau,' *PhaenEx: Journal of Existential and Phenomenological Theory and Culture*, Vol. 2, No. 2 (Fall/Winter 2007), 263.
38 Brown, 'Becoming-Animal in the Flesh,' 261.
39 Deleuze & Guattari, *A Thousand Plateaus*, 314.
40 Rosi Braidotti, *Metamorphoses: Towards a Materialist Theory of Becoming* (London: Polity, 2002), 2.
41 King, *The Shining*, 328.
42 Caillois, 'Mimicry and Legendary Psychasthenia,' 23 & 30.
43 Caillois, 'Mimicry and Legendary Psychasthenia,' 24.
44 Caillois, 'Mimicry and Legendary Psychasthenia,' 28 & 30.
45 Sigmund Freud, 'Medusa's Head,' in *Writings on Art and Literature*, trans. James Strachey (Stanford: Stanford University Press, 1997), 264–265.
46 Steven Connor, 'Fascination, Skin, and Screen,' *Critical Quarterly* Vol. 40, No. 1 (1998), 14.
47 Caillois, 'Mimicry and Legendary Psychasthenia,' 30.
48 Caillois, 'Mimicry and Legendary Psychasthenia,' 30.
49 Sears, *Stephen King's Gothic*, 159.
50 Sigmund Freud, *Leonardo Da Vinci: A Psychosexual Study of an Infantile Reminiscence*, trans. A. A. Brill (New York: Moffat, Yard and Company, 1916), 85.
51 Esther Leslie, 'Introduction,' in *On Photography*, ed. Esther Leslie (London: Reaktion, 2016), 20.
52 Freud, *Leonardo Da Vinci*, 90.
53 Jacques Lacan, *The Seminar of Jacques Lacan: The Ego in Freud's Theory and in the Technique of Psychoanalysis, Book II*, ed. Jacques-Alain Miller, trans. Sylvana Tomaselli (Cambridge, New York, New Rochelle, Melbourne & Sydney: Cambridge University Press, 1988), 50.
54 King, *The Shining*, 328.
55 Caillois, 'Mimicry and Legendary Psychasthenia,' 24.
56 Deleuze & Guattari, *A Thousand Plateaus*, 233.
57 Frank O'Connor, *The Lonely Voice: A Study of the Short Story* (New York: Melville House Publishing, 2004), 29 & 30.
58 Wilcox, 'Drones, Swarms, and Becoming-Insect,' 31.
59 King, *The Shining*, 387 & 234.
60 King, *The Shining*, 310.
61 King, *The Shining*, 300.
62 Chion, *The Voice in Cinema*, 21.
63 Chion, *The Voice in Cinema*, 24.
64 Chion, *The Voice in Cinema*, 3.
65 Chion, *The Voice in Cinema*, 5.
66 Chion, *The Voice in Cinema*, 23.
67 Carlos quoted in Vincent LoBrutto, *Stanley Kubrick: A Biography* (New York, London, Victoria, Toronto & Auckland: Donald I. Fine Books, 1997), 448.
68 Stephan Sperl, *Die Semantisierung der Musik im filmischen Werk Stanley Kubricks* (Würzburg: Könighausen & Neumann, 2006), 179. [My translation. All subsequent translations mine.]
69 Alfred Döblin quoted in Michel Chion, *Sound: An Acoulogical Treatise*, trans. James A. Steintrager (Durham & London: Duke University Press, 2016), 198.
70 Carlos quoted in LoBrutto, *Stanley Kubrick*, 448.
71 King, *The Shining*, 306 & Sperl, *Die Semantisierung der Musik*, 196.
72 Carlos quoted in LoBrutto, *Stanley Kubrick*, 448.
73 Chion, *Sound*, 40.

74 Chion, *Sound*, 40.
75 Luckhurst, *The Shining*, 79.
76 Deleuze & Guattari, *A Thousand Plateaus*, 302.
77 Roger Luckhurst, 'Corridor Dread: A History of Institutional Fear,' *The Wellcome Collection*. Available via: https://soundcloud.com/wellcomecollection/corridor-dread-a-history-of-institutional-fear [Accessed 10 October 2019].
78 Michel Chion, *Audio-Vision: Sound on Screen*, trans. Claudia Gorbman (New York: Columbia University Press, 1994), 75 & 144.
79 King, *The Shining*, 284.
80 Ligeti quoted in Sperl, *Die Semantisierung der Musik*, 186.
81 Sears, *Stephen King's Gothic*, 189.
82 Roger Luckhust @TheProfRog. https://twitter.com/TheProfRog/status/1268438095671832577/photo/1 [Accessed on 5 June 2020].
83 Michel Serres, *The Parasite*, trans. Lawrence R. Schehr (London & Minneapolis: University of Minnesota Press, 2007), 52.
84 King, *The Shining*, 324.

5 Othered Form and Insectile Subjectile

Under the Skin

In Jonathan Glazer's 2013 film *Under the Skin*, four iterations of what I call the insectile, that which pertains to an entomological imagination or fascination and regime of signs, appear.[1] The film, adapted from Michel Faber's novel of the same name, published in 2000, follows an unnamed, extra-terrestrial woman (Scarlett Johansson) engaged in a (forced) labour of extraction, whose purpose is never explained in the film: the objects of the harvest are young, white men, whose flesh or skin meets some unknown demand. The woman serves as bait, intent on luring her victims to the scene of being reaped. A vehicle of male heterosexual fantasy and desire, she is, however, also clearly marked as other. She must be animated, her affective responses—a smile irradiating her face—switched on to perform the part (a white woman, object of desire and 'human' subject) assigned to her. In the final stages of the film, in an unbearable act of violence which culminates in her being burned to the ground, her technological otherness is further revealed as racial difference. Rather than creature of light and glass,[2] she is obdurately black, opaque, a presence that is pulverised.

My argument in this chapter, and in the book more broadly, is motivated by the desire to dismantle the fantasy of the so-called human by being concerned with a fundamental instability at the 'heart' of said subject. This instability is articulated as insectile, deployed with reference to two apparently opposing conditions: form and formlessness or what Georges Bataille calls the *informe*, that which 'does not, in any sense whatever, possess rights, and everywhere gets crushed like a spider or an earthworm'.[3] Explicitly linked to life that is not assigned any value—worm and spider are conceptually aligned with the insectile as rightless—the *informe* 'declassifies' or destructures; it is the 'negation of definition'.[4] Writing about Bataille's term, Yve-Alain Bois observes that he considers the *informe* to be an operation of slippage rather than a theme, substance or concept.[5] At the same time, the insectile, while *informe*, also codes 'armoured life,' even if, as Steven Connor notes, 'the insect as armour in fact is a defence against unrepresentability,'[6] the multiplicity cast out of individualised insect forms. Armoured life is indexed to the fixity of form,

DOI: 10.4324/9781003322757-5

so that both form and formlessness are harboured in the insectile, effectively pulling in these two directions at once. The suffix added at the rear end of the body of the word insect is designed to perform the 'pulsations'[7] between form and the *informe*.

On the one hand, what's at stake in this analysis of *Under the Skin* is an investigation of a racial imaginary or faciality predicated on the insectile. On the other, through the film's insectile sonic events disarticulating the armour and genre of the 'human,' it concerns the unravelling of precisely this logic, that is, the logic of racialisation and the fixity of form. As a critical operation, the insectile interrogates what subjective forms are imposed, kept safe, or rendered expendable in the ways in which they are linked to insects; it might also follow the movements of insects as they appear in films, certain genres, in the functioning and interruptions of genre. (We might want to think, for example, of Manny Farber's description of Claire Denis's films as 'termite art,' because her imagery creeps up on the spectator long after having left the cinema;[8] or the Overlook Hotel in *The Shining* (1980), coded as insectile (see Chapter 4); or the fly in *The Terminator* (1984), settling on the T-800's face without the cyborg flinching or swatting it away: the fly is the marker of the non- or in-human, of a death that is living.)

The insectile functions in a number of ways in *Under the Skin*. There is, for one, the score music, sound events formed of dense crescendo string sounds that establish an extensive and crawling aural field constitutive of the entire fabric of the film. These extra-diegetic, disembodied sounds—no source of emission can be identified—spread everywhere, lose themselves in other mechanical signals, recede into a background buzz or are abruptly cut off, before insidiously emerging once more to, growing in volume, repeat the cycle. Then there are more oblique, marginal occurrences that consist of the close-up of an ant, lifted up to eye level on Johansson's hand having just finished undressing the body of a second woman, her predecessor lying dead at her feet; a fly on a windowpane during a mirror stage scene, aurally and visually breaking in on a moment of self-reflection; finally, the flayed skins of the harvested men—as if they'd undergone ecdysis and thereby allowed another's moulting—floating in a curiously thick, black space.

I want to examine these events to reflect on processes of forming and unforming concerning the subject on- and off-screen, a dynamic to which the film from its opening stages—where noise gradually, more or less, coheres into speech—refers. This dynamic between form and the *informe*, or between figure and ground, plays itself out across multiple systems of representation of the insectile in the film and is, at first, linked to the formation of the face and the absence of expression. I begin, in a section titled 'Face/Form of the Other,' by looking at one of the episodes of insect encounters referred to above (woman/ant) to investigate the mechanisms of (racialised) facialisation in *Under the Skin*. This apparatus of signification, because rendering the face of the woman as arrested and inscrutable, establishes a zone of correspondence with the 'no-face' of an ant. The second part of the chapter, '*Une mouche entre*,' looks at a scene occurring at the usually crucial moment of a mirror stage, on which audience identification with a character often hinges, and gradually moves into the sonic economy of the film on the back, or the buzz, of an interrupting fly. This disturbance occasions a critique of the form of the 'human,' relying on Derrida's work on the law of genre, as well as a shift, during the course of the section, from the on-screen to the off-screen subject, from diegetic to extra-diegetic sound. In this latter part, by lending an ear to the insectile, I seek to respond to Sheryl Vint's claim that the film cannot offer an ethics of difference—she looks elsewhere to find it—as well as to Lucas Hilderbrand's uncomfortable fascination with the film and its, from his perspective, curious, incoherent politics.

While the first section concentrates on form, the rigid systems of racialised othering converging on the face and conceptualised through the insectile, the second part stages an interpretation of *Under the Skin*'s 'anempathetic' score. Michel Chion argues that anempathetic sounds are 'intimately related to cinema's essence,' that is, its 'mechanical nature'.[9] These sounds behave as vectors invoking the cinematic unconscious, in this case also functioning to bury into the filmgoer's repressed 'nature': her mechanical, insectile otherness, in-humanely buried. Chion writes that anempathetic music reveals cinema's 'robotic face,' repurposed here to suggest its insectile 'reality,'[10] a specific kind of filmic body/face and soundscape unconscious, inhabiting or in-forming the subject.[11] This realignment of the anempathetic effect from the robotic to the insectile interprets insects as *technē* and vice versa, while further gesturing towards Greg Hainge's work on sonic cinema.[12] In his book about the French filmmaker Philippe Grandrieux, Hainge proposes sonic cinema as a concept that extends beyond sound, suggestive of an immersive environment into which 'you' fold. He thereby suggests sonic cinema as 'an alternative or perhaps corrective to the idea of haptic cinema' to account for dimensions of sensory encounters no longer structured around the scopic, still the predominant mode of engagement even in the haptic approach.[13]

I want to offer the insectile as a supplement to already established phenomenological ways of approaching a film by structuring my reading according to various aspects of Lacanian psychoanalysis that undergird my analysis. The insistence on the inhuman within explains this chapter's largely latent but committed orientation towards Lacan, bearing in mind his focus on the dehiscent subject, defined by its intimate relationship to the exteriority of the other: the subject internally archives the other. Then there is the process of subject formation which, in Lacanian thought, is structured around mimicry, illuminated by Roger Caillois, whose work on insects and adaptation to space influenced Lacan's development of the mirror stage. Even the word imago, the assemblage of the subject in light of the other, is of entomological origin. Imago is at once fantasy of form, the totality-form of the subject, an armoured self, and indicative of the subject's *méconnaissance* as whole and as 'human'.[14] Post-mirror stage, the subject's knowledge of itself as incomplete, decentred, and fundamentally alienated, is relegated to the unconscious. The body in pieces (or the flayed body)[15] emerges in dreams, but it is also called forth by a cinema 'about' the instability of forms.

The theoretical apparatus I assemble here is constituted through several voices—Lacan, Deleuze and Guattari, Derrida, Michel Serres, and so on—each of which contributes to the economy of the chapter. In one way or another, these voices are all preoccupied with form or the *informe*; I draw them together to think through the dimensions of the so-called human in its insectile dis/articulation. Part of what this assemblage is doing is to highlight the process of writing itself as swarming, always belonging to the other. In writing this piece (and others), I necessarily have to enact the insectile otherness that inhabits me—writing as inhabitation—even if I, a hallucinated subject in command, am still held responsible for shaping it into some kind of form. The insectile is a project about the morphology of the subject, intertwined with entomological events. The topology of the chapter—a rhizome, a teeming multiplicity—folds into insectility, its fabric composed of citations, perhaps experienced as parasitic intrusions. The insectile interrupts the fantasmatic image of the 'human,' whose interiority is always already extimate, Lacan's word for the other, exterior, lodged within. *Under the Skin*, I suggest, testifies to the insectile as radical turmoil of the subject, the extimate intimately disturbing the discourse of form.

Face/Form of the Other

There have been a number of investigations of the film, by Sheryl Vint, Lucas Hilderbrand, Laura Tunbridge, Zara Dinnen and Sam McBean, the last of which is a collaborative piece touching on Scarlett Johansson's underperformance as technological other.[16] Tunbridge's subject, similarly, is Johansson's 'haptic voice' variously deployed or refused in *Under the Skin*, Spike Jonze's *Her* (2013) and Luc Besson's *Lucy* (2014).[17] I want to draw on Dinnen and McBean's work, criticising Ara Osterweil's argument that the film's 'true inquiry is into femininity,'[18] to instead suggest that its interest lies in the face and facial recognition. Johansson, after all, is the 'face' of contemporary SF cinema[19] and therefore 'quasi-object'—because faces always somehow elide the means to capture them—revealing the face as technology and, specifically, as gendered and racialised technology.

According to Deleuze and Guattari, the face is something that is produced by an 'abstract machine of faciality'[20] bestowing subjecthood by recognising only particular elements as being worthy of this signifier. The faciality machine is, hence, a subject-making machine. The organisation of the face defers, in concrete terms, to the 'White-Man face'—the good form—which can't abide alterity, denying the privilege of subjecthood to gendered and racialised others.[21] The face, consequently, is not naked but 'overcoded': it is inscription, a function in a grid which 'rejects faces that do not conform, or seem suspicious'.[22] Insects as 'figures' without face are grasped, in discourses reifying racial difference, as total others; the word 'figure' stands in quotation marks to put it under pressure. Speaking about the fly, Connor notes that it 'does not form a figure, nor yet exactly figure a form,' because it is 'unfigurable,' without face in anthropocentric thought.[23] A figure has a face, is or must be expressive, conform to culturally sanctioned and socially recognised modes of behaviour. The 'faceless' insectile is set to work in fascist and racist-capitalist ideology operating at different levels or scales, conceiving of migrants and refugees in terms of the *informe*, that is, swarms, packs, hordes that raze like locusts (or vermin). In this system of meaning, the insectile is 'figured' as technology of racialised othering, which in *Under the Skin* at first sight functions in contradistinction to the racialising trope of 'animatedness' that Sianne Ngai analyses in *Ugly Feelings* (2007). Underlying this notion of animation is inertness, the activation of a 'lump' dramatizing, or compelled to dramatize, its infusion with life. Not only is the racial stereotype of the silent, inexpressive Asian operative in American culture, but Ngai's reading of John Yau's 'Ghengis Khan' poem cycle (1989–1996) further demonstrates how crucial animation is to the production of the 'overemotional,' racially marked subject.[24] The non-expressive body/face does not disturb racial epistemology but, far from it, precedes and supports it. Impassivity functions as a good only for those whose subject status is never in question, who are not required to prove or enact their 'humanness'. In any other subject, impassivity is suspicious because mechanised, non-human or not quite human, indicative of the secret glitch of unassimilable otherness which, according to this logic, must be effaced, crushed like a spider or earthworm.

The 'White-Man face' organises structures of recognition, to the point of making all others either invisible or hyper-visible. Johansson's face, a.k.a. the cinematic face, is made and unmade by blackness, established as matter of petrochemical technology (oil, plastics) and, at the same time, operating as racialising apparatus intent on securing concrete white faciality. If the face is object rendered technologically, as 'abstract machine' that presents itself as totally transparent, it is imagined as surface to be decoded at first glance. Systems of facial detection produce surfaces or faces as returns: the return of that which is already known and expected, a pattern determined in advance. There are other ways of looking at surfaces, however, as Deleuze demonstrates in *The Logic of Sense* (1969), where he considers Lewis Carroll's *Through the Looking-Glass* (1871). He notes:

> events, radically differing from things, are no longer sought in the depths, but at the surface, in the faint incorporeal mist which escapes from bodies, a film without volume which envelops them, a mirror which reflects them, a chessboard on which they are organised according to plan.[25]

To skirt along the surface, as Catherine Constable shows in her article about Alex Garland's *Ex Machina* (2014), means to be alert to the various constructions and functions (a mist, a mirror, a film) of the surface, continually evolving in Garland's movie. They have the potential, she continues, to 'make us see the limitations of our familiar conceptual categories, such as: humanness, gender, and genre'.[26] Lauren Berlant similarly travels the surface of Gregg Araki's *Mysterious Skin* (2004), thinking about the distribution of life lived at the surface. Berlant is interested in underperformed emotion, flat affect, recessive action. These 'structures of unfeeling' are difficult to decipher, might be a defensive action rather than indicative of a 'casualization of emotion,' not invariably the result of a crisis. Instead, they might be constitutive of a 'space-making device,' thereby establishing a possibility, in the sense that an encounter remains unfinished.[27] There is a politics at work here in terms of, for example, a woman's 'duty' to smile,[28] to yield herself up to codes of behaviour in public, an animation that Johansson, in *Under the Skin*, performs in her seduction game. Berlant's work prompts us to think about the politics of performing emotion (c.f. Ngai) as much as about the difficulty of assessing affectively flat registers. These studies of surface effects help organise the response to the production of the inscrutable other in *Under the Skin*, which appears to be 'all outside,'[29] pure surface, an exoskeletal 'figure' with no inside and without depth.

An 'imploded' face, a dead face, can be the face of late capitalist value,[30] a reading that, despite the conditions under which the woman suffers, would be mistaken in relation to *Under the Skin*. Coded as technological other, the woman's unforthcoming face is explicitly linked to an ant's, a creature she seems to recognise more than the dead woman at her feet.[31] The scene in question takes place in a van, an incongruous interior in terms of both light and dimensions. In this luminous white space, rendering figures as photographic negatives, the scene unfolds as such:

1 a close-up of the dead, clothed, woman, eliciting no response because there is no reverse shot of the naked woman (Johansson) looking down at her predecessor: the expected response, consequently, is experienced as missing. Similarly, no close-up of Johansson precedes the gaze directed at the corpse, so that the shot of the dead woman's face, the first thing we see upon entering the van, occurs, as it were, from nowhere or, more accurately, as if emanating from a sardine can. Lacan, who relates his *point de rencontre* with a sardine can floating on the

water and glimpsed on a fishing boat in *Four Fundamental Concepts of Psychoanalysis* (1973), distinguishes the eye from the gaze, which develops out of objects 'looking' back at the I, in turn displaced as a subject in command of perspective. What brings to mind the sardine can is that the woman's gaze, unmoored as it is from her face, issues from an agency without subject, for which the other, on the ground, has slipped out of relation—slipped, because we assume a relation between the women, one dead, one alive, and looking so much alike.

2 The task of undressing done, her face kept in partial darkness and obscured through a heavy fringe, the woman, now dressed in the corpse's clothes, once more fixes her gaze on the face of the figure below. A single tear, excess moisture stored in the ducts, escapes from the dead woman's eyes as if mourning her own death in the absence of the other woman's lack of 'proper' apprehension. There is no recognising gaze between the one and the other. The living woman, looming over the corpse, in effect passes over her double, whom she replaces. She stays uninvolved, her gaze impersonal.

3 In a long shot, the woman crouches next to the corpse, picks up what turns out to be an ant, though initially it is not identified as such. The woman's movement, after raising herself back up, is to hold her hand up to her face, appearing to inspect the object (or the hand) in the light, rotating it this way and that. It is only then, after two subsequent close-up shots, the second one further magnifying the first, that the face of an ant, framed by moving antennae and legs, becomes radically visible. Despite its singularity, the ant nonetheless gives the impression of swarming in this scene, otherwise so deathly still (occasionally, a low-pitched, hollow hum can be heard). The ant, as Derrida has argued, is a 'microscopic figure of innumerable multiplicity,' a synecdochal form signalling the nest. The ant is also in itself divisible, its genus name, insect, deriving from cut.[32] It is as if an ant could never be spoken of on its own, but only in its prodigious 'crops' of being. Here, it is enlarged to fill the screen, its teeming, excessive mobility, to 'us,' prompting not a turn towards but a recoiling from.

Given that Hilderbrand locates the woman's 'predatory strangeness' in her fascination with the ant, his interpretive act, in fact, performs such an act of flinching. The mandibles, prominent as they are in the shot, suggest that particular perspective, a 'format' in which ants appear as specifically determined epistemological objects. In *Zootechnologien* (2012), Sebastian Vehlken follows the discursive dynamic in which ephemeral collectives figure, or rather are made to figure, hence the usage of the term 'format'. They are formatted according to particular historical, political and technological conditions.[33] In this vein, Hilderbrand's comment on the predatory nature of the woman's behaviour—by that point an intimation; she only scavenges—is at the very least inflected by a discourse already privileging predation, attentive to mandibles that don't release their hold even when detached from the head, instead of, say, an ant's olfactory sense distribution or its muscular memory. In other words, his reading, though without a doubt a response to what is to come in the film, nonetheless participates in a particular rendition of the life of an ant. Yet there are other aspects to draw into the interpretive circumference, which belong to different frameworks or discursive networks. These allow us to approach the insectile through formats other than 'predatory strangeness,' evidently sustaining processes of racialisation that secure the 'White-Man face'. The correlations between representations of insectility and racialisation instrumentalise the insect as absolute other, resonating with the ways in which the matter of blackness is deployed, as Hilderman in fact shows, in *Under the Skin*. This correspondence remains an organising principle which very much lies at the core—dissimulated or blatant—of discussions about the impassive face, itself produced as radical otherness: incoherent, unbearable, unintelligible.

The 'predatory strangeness' that Hilderbrand detects in the woman's face as a result of her interaction with an ant, as such, depends on a tracing that has organised the ant as an aggressive species, an axis of significance that draws on those 'monstrous' mandibles, pincers or shears that can pierce an enemy's skull.[34] Yet tracings, which Deleuze and Guattari understand as something predetermined, 'should always be put back on the map,'[35] itself a thing without form, which can constantly be reworked, unmade. A realignment of perspective—which does not evoke the ant as always already predatory—takes into account the possibility of an affinity, between woman and ant, that can't be crystallised just yet, or not ever. If we place this encounter into the context of the 'supertext'[36] of film and novel—Berlant uses the term to discuss a source text and its other iterations—we might be tempted to find a model of 'kinship' there, but one that invokes a strict order and laws of belonging. In Faber's novel, Isserley beholds sheep and, in their facial features, a correspondence between them based on a shared morphology, which she supposes to be a reason to establish an ethical community. Such a logic is evidently a logic of sameness, which, as Vint has argued, the novel is less interested in criticising, given that it ends on a note of atomised, dispersed indeterminacy: Isserley exploding to become sky, air, particles.[37]

Predominantly focussing on the film, Vint suggests that it refuses to 'humanise' its protagonist and that it keeps insisting on the woman's blackness, captured, by the logic of racialisation, as non-human. She likens Johansson's character and the camera's eye to 'an intellect as "vast and cool and unsympathetic" as that of [H.G.] Wells's [...] Martians,'[38] a comparison occasioned by the impassivity that the film facialises as metaphorically insectile: a face as expressionless as that of an ant. In the supertext proceeding from the novel, the suggestion exists that the gaze, as it passes from woman to ant, supposes a figuration beyond the universe of the film, part of a larger 'whole.' This 'whole' is not a unity but behaves 'like a thread' of communication with other sets or iterations, each of which it prevents from closing.[39] Such a moment of recognition, woman→ant, occurring in the frame of the film, might conceivably lead to further unseen frames, in which the woman = ant. And yet the thread, passing from the novel to and beyond the film, merely returns the ghost of the morphological sameness witnessed in the novel: the logic of the Same is never spelled out in the film. Instead, it exists only as circumstantial outside the frame of the movie, keeping the correlation between woman and ant indeterminate, or impure.

While the woman in *Under the Skin*, as Hilderbrand has noticed, is clearly fascinated by the ant—he conversely acknowledges being fascinated by '[the film's] play with and refusal of conventions of cinematic narrative and representations'[40]—the phenomenon of fascination, so essential to subject formation,[41] tells only half the story. In a scene about an hour into the film, set inside the lair of a ruined house, the woman, in a persistent shot, catches her reflection in a mirror, so often the *Angelpunkt* of an action, rooting her to the spot. The scene cites what functions as a type of Voight-Kampff test the woman was subjected to earlier by her handler/motorcycle man: she now seems to check herself for signs of empathy after the release of a severely disfigured man. Her expression registers no visible change, her face remaining inscrutable. The shot, accompanied by the sound of dripping water, lingers head-on, before the perspective suddenly comes from further away, to the side and slightly to the back. Held there, the woman turns, back towards us, and then we lose sight of her: a fly has entered the frame.

At this point, I want to invoke Derrida's 'The Law of Genre', which informs the remaining section of this chapter. Derrida argues that within the law of genre lies a law of impurity, which means that belonging always relates to non-belonging,[42] that a genus is the 'place' where morphology can't ever really be articulated. The de-structuring of genre applies as much to the literary, poetic and artistic than to gender and racialisation, determining the so-called human; his article allows us to think about impurity as much as the various norms and limits governing a genre. Considering that *Under the Skin* provokes questions relating to the genre of the 'human'—what qualifies a body to be included, or otherwise barred from, appearing in a certain genre—I will show that the fly interrupting the mirror stage scene performs, and is inscribed as, the fundamental generic disturbance of the so-called human.

Une mouche entre: The Sonic Event of a Fly

If the insectile on the face of it and in its faciality is (over)determined as *tout autre*, it is nonetheless both form and formless, read also polymorphous, suggestive of the 'adventure of reading and interpretation' because '[crawling] with thousands of meanings.'[43] The insectile can 'figure' differently, does not necessarily command a form of attention restricted to 'tracings,' something 'ready-made,' having crystallised into a definite form.[44] Consequently, the insectile does not necessarily have to operate according to a politics seeking to reify difference—the ready-made format triggering repulsion or aversion—but allows an orientation towards affective 'forms' that somehow bind, are curiously attuned between bodies.

In this second part of the chapter, the insectile—supplement or suffixed element—latches onto Hainge's concept of sonic cinema, responding, as it does, to dimensions that exceed Laura Marks's concept of haptic visuality and Vivian Sobchack's cinesthetic, embodied subject.[45] Hainge notes his 'discomfort' with the haptic, especially as it develops out of Sobchack's thought, remarking that 'the only possible relation to the cinematic text continues to be figured in terms of visuality, even if the sensory organ in play has changed'.[46] Hainge argues that the haptic fixes in place, fingers knowing what they're looking at, and that as a result it can 'only be felt to vibrate in harmony with each other and with us or [...] instigate a jarring, dissonant relation'.[47] The sonic, by contrast, resonant and constantly re-forming, plays with figure or form and ground and can't be known. Davina Quinlivan, in *The Place of Breath in Cinema* (2012), mentions a 'kind of disparity'[48] taking place between image and sound occupied by the process of breathing, escaping the discourses put forward by Marks and Sobchack. Hainge and Quinlivan identify remainders that are not necessarily or not exclusively material—materiality is troubled in sonic cinema or the 'extra-materiality' of breath and noise[49]—and are unable to be 'grasped' visually.

In the first part of this chapter, the folding of the subject with the insectile occurs in terms of the correlation set up between unnamed woman and insect: the insectile operates diegetically and as form. In this subsequent part, the entanglement subject-insect is constituted extra-diegetically and takes place between cinemagoer and the ground or 'skin' of film. Generative of an enveloping fold,[50] the insectile describes a particular way in which bodies are brought into contact because, adapting Connor's work, the insectile—'too big for space, too packed and too polluting'[51]—is space itself. 'You' fold into this space, the body of the film, encountered as sound, and the film's body in turn folds into 'you'.[52] Because the title of the film, *Under the Skin*, already announces it as a central organ of experience, skin is produced as site of attention: it is an organ formed and unformed through the sonic. Sound evidently belongs to the paradigm of cinema as sensory embodiment or entanglement and possesses haptic and tactile qualities. After all, it is wave phenomenon, proceeding from an object that must be touched, and can be conceptualised as, or at least be related to, skin.[53] It is something felt on, and below, the skin, to the point that sound, skin and space seem coterminous, thereby at once body or form-giving and, as phenomenon of dispersion and propagation, form-destroying.

The link between the insectile and the environment of the skin exists in the broader cultural imaginary. In genre fiction, insects are frequently deployed to render unspeakable trauma, drug abuse or severe psychoses—hallucinations; delusions of invasion—as well as vivid dreams at whose kernel squirm termites or where giant insect eggs are waiting to hatch. Both iterations of *Candyman* (Bernard Rose, 1992 and Nia DaCosta, 2021), William Friedkin's *Bug* (2006) and David Cronenberg's *Naked Lunch* (1991) come to mind, but the imagination of the multiple, of excessive 'life,' the latter unthinking, relentless (like a drive), similarly arises in Donna Tartt's *The Little Friend* (2002). Here, a drug addict's vision, and by extension entire existence, is textured by bugs:

> Points of light, glittery dust flecks like creatures in a microscope—meth bugs, that would be your scientific explanation, because every itch, every goose bump, every microscopic speck and piece of grit that floated across your tired old eyeballs was like a living insect. Knowing the science of it didn't make it any less real. At the end, bugs crawled on every imaginable surface, long, flowing trails that writhed along the grain of the floorboards. Bugs on your skin that you couldn't scrub off, though you scrubbed until your skin was raw. Bugs in your food. Bugs in your lungs, your eyeballs, your very squirming heart.[54]

It is not surprising, as such, that an analysis intent on the audio-visual insectile is affected by and situated within this larger realm. A whole machinery of references has mobilised in the direction of skin and the unconscious 'infested' with insects, across periods, genres, forms, contexts. Insects are space but also the space and skin of the subject. The insectile is that which entangles interiority with the outside, always already existing within, making up the subject in its dehiscence. The pivot on which the section below turns and turns back, then, moves from sound to skin, interior to exterior, form to the *informe*, the inside of the narrative to the cinemagoer losing subjective, humanized, stability. What the extra-diegetic score does is 'proper' alterity back to 'us': the film's insectile body folds into the cinemagoer through sound. From this 'position' or figuration—which Rosi Braidotti identifies as a 'vision of the subject as a dynamic and changing entity'[55]—an 'improper' ethics of difference can begin to be thought.

It is unclear whether the woman subjecting herself to an empathy test in the ruined house during the mirror stage scene to try and determine her 'humanness' is interrupted, riven from her in(tro)spection by the fly, or whether the disturbance is ours or is destined for us alone. Flies are engines of disruption, while, as Connor shows us, they are also indicators of 'a sudden convulsion of scales.' They make 'perspectives collide' between immensities on either side of the windowpane on which the fly alights in the film: grey light outside, dark space inside.[56] The drama in this particular scene concerns the prolonged interval between a reflexive interiority, all the more striking for a character without apparent psychological depth, and an interrupting outside.

The fly, however, agitates not only spatiality but also temporality. Right at the start of one of his books on sound in film, Michel Chion references Victor Hugo's book of poems *L'art d'être grand-père* (1877), whose last line reads: '*Une mouche entre. Souffle immense de la mer.*'[57] A fly arrives, interrupting a soundscape (the sea) that is constant or eternal, an event Chion uses to discuss the comings and goings of sound, the building up of environments instantly destroyed, receding, being moved into the background by random, mean forces.[58] In *Under the Skin*'s mirror stage scene, with its close-up of an impenetrable face swathed in shadows, time is felt yet simultaneously seems suspended. The woman's footsteps, indicative of 'traversing time,' heard just moments before, are arrested, leaving the 'ultramusical' rhythm of dripping water faintly audible.[59] These sounds at once mark 'hourglass time,' a drop-by-drop flow of time evoking time spent,[60] and a temporality experienced as never-ending: the soundscape of the trope of the haunted house, an arrested space where time has ceased to matter. The close-up of the face reinforces the impression of time suspended, immobile as it is, gradually emerging into a dim source of light from a mottled glass window: every other movement has come to a standstill.[61] It is this curious space-time that the fly interrupts: an interiority usually off-limits, if not deemed impossible; a 'mute' faciality that itself appears eternal, because fixed, and is probed in a setting promising, but never really delivering, a structure of understanding.

Given the function of mirrors and the close-ups of faces as affection-images, which remove the face from its spatio-temporal coordinates to, in this case, render it as 'petrified,'[62] the fly breaks our hold on the situation, tenuous as it is. We have already been denied access through the lack of lighting and the camera's movement away from the woman's face. The fly 'swivels' space (inside/outside);[63] it is also liminal creature, not only because it routinely crosses from death into life and vice versa but because its buzz switches between foreground and background, figure and formlessness.[64] It is, in fact, 'aggregate,' adopting Leibniz's terms to refer to multiplicities, not a 'well-formed object' but 'irrational,' something nebulous. Undefined, as such, by the concept of the border, it is *Ungestalt*, its buzz akin to the 'basic element,' according to Michel Serres, of our logos.[65] The fly points, then, to this 'ground' of being, with which it merges, into which it disappears, and which sounds inside 'us':

> [Noise/Buzz] settles in subjects as well as in objects, in hearing as well as in space, in the observers as well as in the observed, it moves through the means and the tools of observation, whether material or logical, hardware or software [...].[66]

The fly's buzz is itself everywhere and ceaseless, only occasionally perceptive as phenomenon, when it is in effect event, ongoing, eternal. In *Under the Skin*, the fly, even though operating as phenomenon in this scene, serves to draw our attention to the event of restless disruption that un/forms the subject, particularly as it (mis)recognises itself as 'human.'

The film defamiliarises the 'human'; as Vint writes, it 'probes the limit of how humans see one another.'[67] Hilderbrand adds that it 'works to structurally frustrate identification with the film's protagonist,' and refers to the scene discussed above, 'keeping her a seeming other.'[68] What happens here is that the fly disarticulates the 'human,' like it fundamentally does in Cronenberg's *The Fly* (1986). Flies are closely associated with the 'human' all the while so totally, intimately, other: they are 'our' eternal companions.[69] The fly disrupts a mirror stage, apparently motivated by the question of the 'place' of the 'human' (where does 'humanness' reside?), though disruption, as Lacan tells us, is integral to the process of the mirror stage. The investigative gaze into the mirror does not simply belong to the reflected subject in this film but is ours, too. We are trying to read a face and are prevented from doing so. The fly is irritant as well as vector of a lack of recognition. A reminder of how inscrutable the woman is, the irritation caused by the fly, however, goes further than that. Its buzz, intermingling with the sounds of dripping water, enervates before the source is seen and grates the skin, the matter, 'envelope' or 'milieu' of our own subjectivities.[70]

The inspecting gaze is a policing gaze; the film's citation of the Voight-Kampff test makes that much clear, and it effectively puts us into the position of the bounty hunter Deckard (in more ways than one, 'we' are replicant). At issue is the genre of the 'human,' understood and produced as white—the marker, in an anti-black world, of what it means to be considered 'human'—but the fly is also that which doesn't respect borders or norms. As I have tried to show, the interplay between form and the *informe*—a system of racialised alterity on the one hand, initiating the disarticulation of the so-called human on the other—is indexed to the insectile in *Under the Skin*. What, then, if we were to conceive of this fly as the 'principle of contamination' right at the heart of the law of genre,[71] that speck or trait which structures and at the same time undoes everything? Derrida's vocabulary certainly suggests as much. In an aggregation of terms, the 'essential disruption' of the law of genre also goes by the following names: 'impurity, corruption, contamination, decomposition, perversion, deformation, even cancerization, generous proliferation, or degenerescence'.[72] Flies are associated with most, if not all, of the above, as carriers of disease, purely libidinal creatures, angels or 'anti-angels'[73] of death and decay. They are, Connor observes, 'the embodiment of spatial and categorical disturbance,'[74] to which, proceeding from the reading above, we might add ontological disturbance too.

The fly is a 'figure' which, in the vicinity of the mirror, contributes to frustrate identification (filmgoer→woman) but, even more significantly, is visual and aural 'image' breaking apart the hallucinated unity of the subject as fully 'human.' The reason it does so is because here, and throughout the film, an insectile buzzing occasions an itching skin, as if 'we' were the host to millions of swarming things, penetrating to, and living in, 'our' squirming hearts. This embodiment or memory of insectile dehiscence, I suggest, means that the film is, after all, able to move toward an ethics of difference, which Vint thinks it cannot do. Indeed, if we stop short of handling the film as an engagement with sound and/as the skin of the cinemagoer, she is right to argue so, considering that final moment of xenophobic violence, leaving no means from which to depart in search of such an ethics. Narratively, or purely in terms of the scopic, that is, the film can't provide any alternative encounters with the other/ed, absolutely and disastrously expelled. The scene in question shatters, is shattering, and then the film ends: it gives the impression of having reached a limit (for Vint, the limit of an ethics of sameness). I want to keep insisting on the sonic, however, reverberating, breaching limits: the ethical, like justice, similarly is that which lies outwith the law, beyond the horizon.[75]

Cinema, according to Jennifer Barker, 'entails a whole range of possibilities of touch against our skin: films can pierce, pummel, push, palpate, and strike us; they also slide, puff, flutter, flay, and cascade along our skin.'[76] In this instance, the narrative already insists on the politics of the skin. The skin is that which is extracted from straight white men, expendable because, although unspoken, the issue of class remains latent, is to be found in the geography of the city where the film is set. The woman looks for solitary men and generally goes on the prowl in deprived areas in Glasgow, around Ibrox and Parkhead—the first victim whose abduction we witness is a Celtic supporter—on Trongate, lower-end to the adjacent Buchanan Street, Glasgow's high street, and declining in its approach to Glasgow Cross and Gallowgate, as well as the vicinity around Glasgow airport, all of which are regions blighted by poverty and crime. During one entrapment scene, the camera follows the victim down, into a 'sunken place,' to borrow the term from Jordan Peele's *Get Out* (2017), a space like cast resin. Here, he witnesses the 'puff' of an amorphous, desiccated man, leaving only the veil of his floating skin. Skin is that which gives the body form. It is, as it were, the 'body's face, the face of its bodiliness,' as Connor puts it,[77] and figures the 'human'. The skin of white men is that which is required to pass, providing the means for a transformation or metamorphosis into a non-racialised, therefore 'human,' subject.

At the same time that the film is preoccupied with the form of the skin or the skin as form-giving, however, it undoes the function of skin as that which holds together, its capacity to possess and preserve form and figuration. It does so through that recurring insectile sonic presence, whose provenance can't effectively be pinned to a source, emanating from an un-bodied or many-bodied predatory gaze, and constituting an aural field that agitates and itches. It stays enigma throughout, the universe it creates at once external (coded as total outside) and internal, because transposing its logic of the outside onto 'my' skin. This sound makes the skin function or feel like the concealed and extimate environment of the unconscious: I am, in fact, other. The extensive sonic environment in/of the film produces an impression of 'my' skin not as colander from which things leak[78] but as aggregate, host, invaded surface and evaded interiority, that is, the repressed reality of a dehiscent subject.

The skin, according to Serres, is 'milieu,' which, so Connor explains,

> requires a physics of the imagination that lies between the conditions of liquid and solid. The implicative capacity of the skin—its capacity to be folded in upon itself—means that it is involved in other, much more mobile and ambivalent substances too, substances and forms which do not have simple superficiality or absolute homogeneity, but in which, so to speak, the surface turns on itself, goes all the way down: smoke; clouds; dust; sand; foam.[79]

Connor is interested in those 'moments of umbilical inclension and involution,' when skin suddenly and intimately becomes other, when the outside reveals itself as inside, and such distinctions break apart, are abandoned, or fall away.[80] An 'entire environment' rather than 'surface, membrane, or interface,'[81] the skin further binds thinking to it: thought, so Didier Anzieu, is matter of the skin[82] and therefore of touch. As Claudia Benthien notes, there is an 'epistemological equating between skin and touch or the relationship between skin and hand.'[83] In this vein, the scene analysed earlier, of the woman's rotating hand—holding it up to the light as if inspecting it—already prepares us to think about hand/skin and the touch of the insect as well as about the hand as appendage that doesn't quite belong.

The film's process of estrangement extends to skin, even if its whiteness is used to signify the 'human,' because its figuration, its capacity to figure, is undone through cascading extra-diegetic sounds. A viola, distorted through speed,[84] is hurried along in an 'uncomfortable,' discordant pitch.[85] In interviews, Mica Levi, the score composer, frequently talking about her 'immersion' in the film, elaborates on her work:

> A lot of the sound [in *Under the Skin*] is a mixture of bad recording technique, on my part, and not-fine playing. Violas are so harmonic because they contain a lot of air. A viola is not solid, the sound it produces is like a photocopy of a photocopy of a photocopy of something, because you get an airiness, and creepiness, and there's a struggle in that. The vibrato doesn't ring out. It's dead.[86]

Photocopies of photocopies introduce those 'mobile and ambivalent substances,' such as dust, grains, and so on, in turn producing little 'microrhythms'[87] in the 'image' of a sound. This aural event generates a swarming movement, a busy temporality that unnerves. Even though Levi attributes it to the woman's stomach or hunger,[88] a synch point—the moment of coincidence between sound and source—is, as mentioned earlier, never confirmed. The sound wanders too much, appears to derive from a growing nimbus of light at the start of the film and, a little later, to arrive with motorcycle man. It keeps arising without origin, a deterritorialised sound designed to affect the cinemagoer.

In *Under the Skin*, the subject's corporeal engagement with the opaque matter of its own alienation is determined by sound, the swarming unseen insect-things that make the skin itch. An itch undoes singularity, the hallucination of coherence. It is, Connor writes, 'the experience of displacement, of the dislocation of the organism from itself.'[89] It is in this proposition that the itch, if you wish, of an ethics of difference is at work in the film. Cinema, according to Deleuze, because of its 'lack,' cannot give us the 'presence of the body,' which, he continues,

> is perhaps also because it sets itself a different objective; it spreads an 'experimental night' or a white space over us; it works with 'dancing seeds' and a 'luminous dust;' it affects the visible with a fundamental disturbance, and the world with a suspension, which contradicts all natural perception. What it produces in this way is the genesis of an 'unknown body' which we have in the back of our heads, like the unthought in thought, the birth of the visible which is still hidden from view.[90]

Seeds and dust are the microrhythms of the audiovisual image, whose 'absence' is generative: an unknown, unthought body incubates. Marked by the itch of the other, the cinemagoer apprehends the multiplicity of her skin,[91] which becomes host: she experiences herself as host. In this sense, *Under the Skin* fundamentally disrupts the iterative structure of the subject as 'human,' all the while spreading an experimental night—a caesura—over the 'naturally' occurring assignation of that appellation. The term 'human' is something that must be bestowed, stolen, from elsewhere. The film's *force de rupture* exists in its insectile disarticulation of 'my' skin, dissolving the humanized form of the subject. In 'Eating Well,' Derrida proposes to 'rearrange' the subject so that it 'no longer dominates from the centre' and suggests describing it as 'subjectile' to record the *jet* of *différance* that exists at the heart of the subject.[92] In light of the analysis above, we might begin to notate the subject as insectile subjectile, marking it in its specifically insectile dehiscence and dimensions.

Notes

1 Jonathan Glazer (dir.), *Under the Skin* (StudioCanal, 2013).
2 See Charlotte Constable, 'Surfaces of Science Fiction: Enacting Gender and "Humanness" in *Ex Machina*,' *Film-Philosophy*, Vol. 2, No. 22 (2018), 292.

See also Donna Haraway, *Simians, Cyborgs and Women: The Reinvention of Nature* (New York: Routledge, 1991), 153.
3 Georges Bataille, *Encyclopaedia Acephalica*, trans. Iain White (London: Atlas Press, 1995), 51–52.
4 Alastair Brotchie, 'Introduction,' in *Encyclopaedia Acephalica: Comprising the Critical Dictionary & Related Texts*, trans. Iain White (London: Atlas Press, 1995), 23.
5 Yve-Alain Bois, 'The Use Value of "Formless,"' in *Formless: A User's Guide*, ed. Yve-Alain Bois & Rosalind E. Krauss (New York: Zone Books, 1997), 15.
6 Steven Connor, 'As Entomate as Intimate Can Pinchably Be,' accessed via: http://stevenconnor.com/insects.html [Accessed 28 July 2020].
7 Bois, 'The Use Value of "Formless,"' 31.
8 Manny Farber cited in Charles Bramesco, '*High Life* Review—Orgasmic Brilliance in Deepest Space with Robert Pattinson,' *The Guardian*, 10 Sept 2018, accessed via: https://www.theguardian.com/film/2018/sep/10/high-life-review-robert-pattinson-claire-denis-sci-fi-drama-astronauts [Accessed on 28 September 2021].
9 Michel Chion, *Audio-Vision: Sound on Screen*, trans. Claudia Gorbman (New York: Columbia University Press, 1994), 8.
10 Chion, *Audio-Vision*, 9.
11 Jacques Lacan, *Four Fundamental Concepts of Psychoanalysis*, trans. A. Sheridan (New York & London: W.W. Norton & Co., 1998), 26.
12 Jussi Parikka, *Insect Media: An Archaeology of Animals and Technology* (Minneapolis & London: University of Minnesota Press, 2010), xiii.
13 Greg Hainge, *Philippe Grandrieux: Sonic Cinema* (New York, London, Oxford, New Delhi & Sydney: Bloomsbury, 2017), 13, 77 & 80.
14 Jacques Lacan, 'The Mirror Stage as Formative Experience of the I as Revealed in Psychoanalytic Experience,' in *Écrits: A Selection*, trans. Alan Sheridan (London & New York: Routledge, 2001), 3.
15 Didier Anzieu, *The Skin Ego*, trans. Chris Turner (New Haven & London: Yale University Press, 1989), 41.
16 Zara Dinnen & Sam McBean, 'The Face as Technology,' *New Formations*, No. 93, Summer 2017, 128.
17 Laura Tunbridge, 'Scarlett Johansson's Body and the Materialisation of the Voice,' *Twentieth-Century Music*, Vol. 13, No. 1 (2016), 145.
18 Ara Osterweil, '*Under the Skin*: The Perils of Becoming Female,' *Film Quarterly*, Vol. 67, No. 4 (Summer 2014), 44.
19 Dinnen & McBean, 'Face as Technology,' 127.
20 Gilles Deleuze & Félix Guattari, *A Thousand Plateaus: Capitalism and Schizophrenia*, trans. Brian Massumi (London & New York: Continuum, 2004), 187.
21 Deleuze & Guattari, *A Thousand Plateaus*, 197.
22 Deleuze & Guattari, *A Thousand Plateaus*, 189 & 197. On the naked, destitute face, see, for example, Emmanuel Levinas, *Ethics and Infinity*, trans. Richard A. Cohen (Pittsburgh: Duquesne University Press, 1985), 86.
23 Steven Connor, *Fly* (London: Reaktion, 2006), 31 & 32.
24 Sianne Ngai, *Ugly Feelings* (Cambridge, MA & London: Harvard University Press, 2007), 94.
25 Gilles Deleuze, *The Logic of Sense*, trans. Mark Lester & Charles Stivale (London: Athlone Press, 1990), 9–10.
26 Constable, 'Surfaces of Science Fiction: Enacting Gender and "Humanness,"' 300.
27 Lauren Berlant, 'Structures of Unfeeling: *Mysterious Skin*,' *International Journal of Politics, Culture & Society*, Vol. 28, No. 3 (2015), 194, 195 & 207.

28 See Fabienne Collignon, 'The Politics of a Smile,' *new formations*, Vol. 95, Jan 2019, 87–102.
29 Connor, 'As Entomate as Intimate Can Pinchably Be'.
30 Berlant, 'Strcutures of Unfeeling,' 198.
31 Lucas Hilderbrand, 'On the Matter of Blackness in *Under the Skin*,' *Jump Cut: A Review of Contemporary Media*. Accessed via: https://www.ejumpcut.org/archive/jc57.2016/-HilderbrandUnderSkin/index.html [Accessed 10 October 2019].
32 Jacques Derrida, 'Fourmis,' in *Hélène Cixous Rootprints: Memories of Life & Writing*, ed. Hélène Cixous & Mireille Calle-Gruber, trans. Eric Prenowitz (London & New York: Routledge, 1997), 120 & 121.
33 Sebastian Vehlken, *Zootechnologien: Eine Mediengeschichte der Schwarmforschung* (Zürich: diaphanes, 2012).
34 I'm relying on Maurice Maeterlinck's *The Life of an Ant*, here. Trans. Bernard Miall (London: George Allen & Unwin Ltd., 1941), 40.
35 Deleuze & Guattari, *A Thousand Plateaus*, 14.
36 Berlant, 'Structures of Unfeeling,' 192.
37 There is another argument to be made here. The novel, insofar as it challenges this logic of sameness, does so by way of class, the notice it takes of the immensely suffering body and standing in such stark contrast to the body of the capitalist (in this case Amlis Vess), which, as Elaine Scarry observes, is a body of non-participation, of being exempt. Under these circumstances, what Scarry further calls the 'magnified body,' a body altered and ravaged by labour, it is unsurprising that the imagination at work would be enthralled by an un/becoming, a losing of form, which capital and its production processes impose. While the utter destruction of the body, with which both iterations conclude, is the result of 'our' catastrophic inhospitability to the other in the film, it is, in the novel, the point when world-annihilating pain can finally come to a stop. See Elaine Scarry, *The Body in Pain: The Making and Unmaking of the World* (Oxford: Oxford University Press, 1985), 265–267.
38 Sheryl Vint, 'Skin Deep: Alienation in *Under the Skin*,' *Extrapolation*, Vol. 56, No. 1 (Spring 2015), 9.
39 Gilles Deleuze, *Cinema I: The Movement Image*, trans. Hugh Tomlinson & Barbara Habberjam (Minneapolis: University of Minnesota Press, 1997), 16–17.
40 Hilderbrand, 'On the Matter of Blackness'.
41 Jacques Lacan, *The Ego in Freud's Theory and in the Technique of Psychoanalysis: The Seminar of Jacques Lacan, Book II*, ed. Jacques-Alain Miller, trans. Sylvana Tomaselli (Cambridge, New York, New Rochelle, Melbourne & Sydney: Cambridge University Press, 1988), 50.
42 Jacques Derrida, 'The Law of Genre,' trans. Avital Ronell, *Critical Inquiry*, Vol. 7, No. 1 (Autumn 1980), 65.
43 Derrida, 'Fourmis,' 119.
44 Deleuze & Guattari, *A Thousand Plateaus*, 13.
45 Laura Marks, The *Skin of the Film: Intercultural Cinema, Embodiment and the Senses* (Durham, NC: Duke University Press, 2000); Vivian Sobchack, *Carnal Thoughts: Embodiment and Moving Image Culture* (Berkeley: University of California Press, 2004).
46 Hainge, *Philippe Grandrieux*, 80.
47 Hainge, *Philippe Grandrieux*, 80 & 81.
48 Davinia Quinlivan, *The Place of Breath in Cinema* (Edinburgh: Edinburgh University Press, 2012), 21.
49 Quinlivan, *The Place of Breath in Cinema*, 26.

50 On this note, see also Saige Walton, *Cinema's Baroque Flesh: Film, Phenomenology and the Art of Entanglement* (Amsterdam: Amsterdam University Press, 2016).
51 Connor, 'As Entomate as Intimate Can Pinchably Be,' n.p.
52 Hainge, *Philippe Grandrieux*, 77.
53 Thomas Elsaesser & Malte Hagener, *Film Theory: An Introduction through the Senses* (New York & London: Routledge, 2010), 131 & 137.
54 Donna Tartt, *The Little Friend* (London: Bloomsbury, 2002), 440–441.
55 Rosi Braidotti, *Metamorphoses: Towards a Materialist Theory of Becoming* (London: Polity, 2002), 2
56 Steven Connor, 'Making Flies Mean Something,' Lecture given at *Beckett and Animality* Conference, University of Reading, 26 September 2009. Accessible via: http://stevenconnor.com/flymean.html [Accessed 10 October 2019], 3.
57 Michel Chion, *Sound: An Acoulogical Treatise*, trans. James A. Steintrager (Durham & London: Duke UP, 2016), 3–5.
58 Chion, *Sound*, 6.
59 Chion, *Sound*, 40.
60 Chion, *Sound*, 41.
61 See Deleuze, *Cinema I*, 87.
62 Deleuze, *Cinema I*, 90.
63 Chion, *Sound*, 7.
64 Connor, 'Flies', 7.
65 Serres, *Genesis*, 2 & 7.
66 Serres, *Genesis*, 13.
67 Vint, 'Skin Deep,' 6.
68 Hilderbrand, 'On the Matter of Blackness'.
69 Connor, 'Flies,' 5.
70 Envelope is Didier Anzieu's word; milieu Michel Serres'. See Anzieu, *The Skin Ego*, trans. Chris Turner (New Haven & London: Yale University Press, 1989). Serres' comes from *Les Cinq Sens*, cited in Steven Connor, *The Book of Skin* (London: Reaktion, 2004), 26.
71 Derrida, 'The Law of Genre,' 57.
72 Derrida, 'The Law of Genre,' 57.
73 Connor, *Fly*, 15.
74 Connor, *Fly*, 15.
75 See Jacques Derrida, 'Force of Law: The "Mystical Foundation of Authority",' in *Deconstruction and the Possibility of Justice*, ed. Drucilla Cornell, Michael Rosenfeld & David Gray Carlson (New York & London: Routledge, 1992), 3–67.
76 Jennifer M. Barker, *The Tactile Eye: Touch and the Cinematic Experience* (Berkeley, Los Angeles & London: University of California Press, 2009), 36.
77 Steven Connor, *The Book of Skin* (London: Reaktion, 2004), 29.
78 Anzieu, *The Skin Ego*, 102.
79 Connor, *The Book of Skin*, 40.
80 Connor, *The Book of Skin*, 39.
81 Connor, *The Book of Skin*, 28.
82 Anzieu, *The Skin Ego*, 9.
83 Claudia Benthien, *Skin: On the Cultural Border between Self and the World*, trans. Thomas Dunlap (New York: Columbia University Press, 2002), 185.
84 Ryan Lattanzio, 'Mica Levi on why composing *Under the Skin* was "really mental",' *Indie Wire*, 10 November 2014. https://www.indiewire.com/2014/11/mica-levi-on-why-composing-under-the-skin-was-really-mental-190232/ [Accessed 10 October 2019].

85 'Interview with Mica Levi,' *The Guardian*, 15 March 2014. https://www.theguardian.com/music/2014/mar/15/mica-levi-under-the-skin-soundtrack [Accessed 10 October 2019].
86 Ryan Lattanzio, 'Mica Levi,' https://www.indiewire.com/2014/11/mica-levi-on-why-composing-under-the-skin-was-really-mental-190232/ [Accessed 10 October 2019].
87 Michel Chion, *Audio-Vision: Sound on Screen*, trans. Claudia Gorbman (New York: Columbia University Press, 1994), 16.
88 Ryan Lattanzio, 'Mica Levi,' https://www.indiewire.com/2014/11/mica-levi-on-why-composing-under-the-skin-was-really-mental-190232/ [Accessed 10 October 2019].
89 Connor, *Book of Skin*, 249.
90 Gilles Deleuze, *Cinema II: The Time-Image*, trans. Hugh Tomlinson & Robert Galeta (Minneapolis: University of Minnesota Press, 1989), 201.
91 Connor, *Book of Skin*, 234.
92 Jacques Derrida, *Points ...: Interviews, 1974–1994*, ed. Elisabeth Weber, trans. Peggy Kamuf & others (Stanford: Stanford University Press, 1995), 268 & 260.

6 Relations of the Third Kind

In *The Space of Literature* (1955), Maurice Blanchot notes:

> The writer seems to be the master of his [sic] pen; he can become capable of great mastery over words and over what he wants to make them express. But his mastery only succeeds in putting him, keeping him in contact with the fundamental passivity where the word, no longer anything but its appearance—the shadow of a word—never can be mastered or even grasped. It remains the ungraspable which is also unreleasable: the indecisive moment of fascination.[1]

For Blanchot, writing initiates and occurs in a state of passivity in which the writer, despite her apparent mastery, yields any grasp of the word. The work is impersonal, having subsumed the writer into its folds: to write means to give up the I. His account of writing gestures towards Freud and his critique of presence; even if Freud is mentioned only once in *The Space of Literature*, psychoanalysis and the kind of speech it allows—always in danger of suppressing the movements of that other language—permeate the scene of Blanchot's writings. If writing belongs to anyone, he says, it is the property of the other, who nonetheless disappears, along with the I, in the concept of a language that 'no one speaks,' which cannot be attributed to any subject in command.[2] Rather, it seems a swarming thing, without centre, a 'giant murmuring,' an impression Blanchot develops in relation to Kafka, in whose writings the I substitutes for another.[3] To write means delivering the I up to a language which destroys the very possibility of ownership and entails a metamorphosis of sorts: to become no one, to be absorbed into a 'neutral force, formless and bereft of any destiny'.[4] That force or space of the neutral is a realm of fascination, defined by an absence of time in which the personal, along with the 'proper,' is erased to make way for the anonymous and faceless (even if fascination tends to be read in the face, prone to a certain slackness, just as the rest of the body stiffens up in order to ward off the threat of fascination).

DOI: 10.4324/9781003322757-6

My argument in this chapter is concerned with writing, or language more generally, and the insectile, the extent to which writing establishes a scene of fascination linked to insects, as a process (including that of speech and of reading) that bears the traces of an entomological fascination and that, as a result, is an 'enterprise in desubjectification,' generating what Deleuze and Guattari call the 'minimal subject' or what I also identify as the insectile subjectile.[5] The chapter is, in a first instance, indexed to Blanchot's work on fascination and the movement of writing in *The Space of Literature* while further turning to consider *The Infinite Conversation* (1969), in which Blanchot speaks of writing as that which 'leads us to sense a relation which is entirely other, a relation of the third kind'.[6] Across the two texts, Blanchot investigates the encounter with the other, which can really take place only when the subject 'surrenders' to the 'indeterminate milieu of fascination'.[7] This milieu is aligned with his concepts of the outside and, as already mentioned, the neutral, the featureless and formless, an 'impersonal' and 'immense' presence seizing the subject, compelling it to 'pass from the first person to the third person, so that what happens to me happens to no one'.[8]

In the texts I consider here, writing is the space where the subject is deformed, pushed into relations of the third kind; these relations, in turn, are staged in terms of the insect other. I begin with Vladimir Nabokov's *Ada or Ardor* (1969) and proceed through several other encounters with insects—A.S. Byatt's 'Morpho Eugenia' (1992), Jonathan Lethem's *Motherless Brooklyn* (2000), to name a few—before arriving at Clarice Lispector's 1964 novel *The Passion According to G.H.* In each of these texts, the enigma of fascination implicates the subject in the insectile, and the scene of writing taking place constitutes itself as a scene of intimacy—or extimacy, using Lacan's term—with the insect other that fundamentally disrupts the being of the subject. In what follows, I will trace relations of the third kind and, through what Derrida calls the *fourmillement* of language, the potential of the I transforming into a minimal subject: insectile subjectile. In the first part, 'Errant Speech,' I focus on the movements between relations of the third kind and psychoanalysis, gradually edging its way towards the insectile. The second part, 'Entomological Fascination,' revolves around Nabokov's *Ada*, while the subsequent section, 'Swarm-Being, Language-Nests,' takes as its object Derrida's 'Fourmis,' put to work, amongst others, with Byatt and Lethem's writings, before arriving, in the final annulation, at Lispector's novel. The structure of the chapter implies a progression of sorts (itself at odds with the concept of becoming), considering that my argument participates in efforts trying to imagine a non-unitary subjectivity citing the insectile as paradigm for a new subjective 'form'.[9] Nonetheless, despite its position at the end of this chapter, Lispector's *Passion* remains a much more ambivalent text than the chapter's structure suggests, ultimately saying as much about the privileged 'human' form than it does about its decomposition.

Errant Speech

In *The Infinite Conversation*, Blanchot speaks about relations of the third kind in terms of a 'man without horizon,' a condition or 'being' to be interpreted not as narcissism—an ego so illimited and fragile that it seeks to incorporate everything into its structure—but rather as the relinquishing of sovereignty. Phallogocentrism stays lodged in the word 'man' and the figure of the writer invariably remains male and is otherwise unmarked and therefore white, able-bodied, bourgeois, and so on. The scene of dispossession as radical possibility is, as such, tied to a particular subject, whose status is never in question and who, accordingly, is free to imagine deprivation, the loss of self-ownership and self-determination instead of having to suffer it. On the other hand, it is precisely that subject which needs to be 'disarmed,' as Susan Hanson puts it in the foreword to *The Infinite Conversation*, so that this act of assumption on Blanchot's part can also be read as the attribution of an urgent deconstructive duty to the phallic subject. That duty is itself without horizon, not least with regard to the injunctions, demanding unceasing contestation and radical openness toward the other, driving his work.[10] This 'man without horizon' can be defined as a becoming-other, an 'involution,' to use Deleuzo-Guattarian vocabulary: there is no regression or progression, no direction to go, no origin to return to, no imago to attain. Becoming-other dispenses with the 'law of the same,' the first relation that takes place in any encounter between the I and the other, which must render subject and object identical at all costs.[11] The second relation similarly preserves sameness or the drive towards unity or the 'rigour' of One;[12] in this case, the other might be affirmed as subject and therefore is granted certain rights (such as dignity and autonomy), but the phallic subject persists as an organising principle and normative term. In other words, I recognise the other as kindred, the same as me. It is only the relation of the third kind that breaks the logic of One, of the phallic subject and of the Same, venturing towards another 'form of speech,' other possibilities of engagement that depart from returning me either to myself or to the figure of an otherwise coherent, self-same I.[13]

There are parallels to be traced between this third relation and psychoanalysis, as Blanchot makes clear in 'The Analysis of Speech,' an essay included in *The Infinite Conversation* and revised from an earlier piece on Freud written in 1956. The relation between the armchair and the couch, analyst and analysand, 'in a space that is separate, cut off from the world,' he observes, occurs, at its best, in terms of an absence of relation.[14] To paraphrase Lacan, whom Blanchot cites at various points throughout his essay: at the most fundamental level, the analysand, beginning analysis, speaks to the analyst as other, someone else, about someone other than herself. Considering that the matter under discussion is always other than what it appears to be—a present conflict stemming from a much older trauma; the defence mechanisms dissimulating the old drama—and that the subject forms through the 'exigency of [...] lack,' the relation that develops (or should develop) between analyst and analysand is structured around absence.[15] The narrative that constitutes itself in this space between couch and armchair—Blanchot further takes into account the phenomenon of transference, 'equivalent to the fascination proper to hypnosis'—corresponds to a 'new invention of language,' namely the language of an 'ungraspable event—ungraspable because it is always missed, a lack in relation to itself'.[16] At the core of the psychoanalytic process is, hence, an impossibility of speech, so that the dialogue that becomes possible, on the basis of this impossibility and in this setting, is one which surrenders to passivity. The two interlocutors, suspended in this space, talking to each other without faces—in the Freudian method, analysand and analyst cannot lock eyes or read each other's facial expressions and thereby are able to give in to their unconscious thoughts—engage in this errant dialogue, a relation of the third kind, in which the fascinated, displaced I speaks 'by way of the other'.[17]

In *The Architecture of Psychoanalysis* (2017), Jane Rendell considers the consulting room (though more often than not still subject to race and class privilege) as a transitional space, like a type of corridor, situated neither inside nor outside. She refers, amongst others, to André Green, working in both the Freudian and Winnicottian traditions, who describes it as an 'intermediate' space, by its very arrangement facilitating the movements of the new language generating within it, arising at its curious limit.[18] Even though Blanchot does not provide any detailed descriptions of the psychoanalytic setting beyond the reference to its enclosure set apart, as an expression of the suspension of the usual rules of communication repressing the intrusions of the unconscious, Foucault, for his part, is attentive to the spatialities in 'almost all' of Blanchot's other writings. He comments on the 'profound' engagement with 'houses, hallways, doors and rooms' as

> placeless places, beckoning thresholds, closed, forbidden spaces that are nevertheless exposed to the winds, hallways fanned by doors that open rooms for unbearable encounters and create gulfs between them across which voices cannot carry and that even muffle cries; corridors leading to more corridors where the night resounds, beyond sleep, with the smothered voices of those who speak, with the cough of the sick, with the wails of the dying, with the suspended breath of those who ceaselessly cease living; a long and narrow room, like a tunnel, in which approach and distance—the approach of forgetting; the distance of waiting—draw near to one another and unendingly move apart.[19]

The context, here, is the correlation between threshold spaces—internal or subterranean passages; waiting rooms and other areas of abeyance;

zones of exposure or precarity—and a speech turned toward the outside, away from the commanding I. The language arising in the analytical situation can, as such, be conceptualised spatially; this spatial imagination similarly informs Blanchot's representation of relations of the third kind. André Green, in fact, calls the relation between analysand and analyst the 'third element,' the 'analytic object,' corresponding to Winnicott's transitional object, situated between the two people involved.[20] It is difficult to banish the resonances, the 'intrusive ideas' and 'side-issues,'[21] on this note, not least considering the focus—the scene of fascination and writing—of this chapter. Freud instructs his fellow analysts and patients to think of treatment as a motorised propulsion through a landscape, as if the patient were travelling by train and describing what they saw to someone else in the carriage.[22] We might also want to think of H.D.'s *Tribute to Freud* (1943), attentive to the consulting room's specificity, the dark passages and hallways beyond its numerous doors. H.D. mentions the room's silence, the sensation of being 'quite alone,' with no sound of traffic disturbing the space from the street below; she is aware of the room as node in a nexus of connections to other areas of the house and feels herself to be the wrong size, too tall for the couch, leading her to think about her taller, older brother.[23] Her brother's size, in turn, prompts her to remember the size of a log which was much too heavy for both of them to move but which their even taller (grown-up) half-brother Eric can shift:

> There was a variety of entertaining exhibits; small things like ants moved very quickly; they raced frantically around but always returned to the same ridge of damp earth or tiny lump of loam. In neatly sliced runnels, some white, wingless creatures lay curled. The base of the log had been the roof of a series of little pockets or neat open graves, rather like Aztec or Egyptian burial-chambers, but I did not know that. These curled, white slugs were unborn things. They were repulsive enough, like unlanced boils. Or it is possible that they were not essentially repulsive—they might be cocoonless larvae, they might 'hatch' sometime. But I only saw them; I did not know what they were or what they might portend. My brother and I stood spellbound before this disclosure. Eric watched the frantic circling of the ants attentively. Then he set the log back carefully, so as to crush as few of the beasts as possible, so as to restore, if possible, the protective roof over the heads of the little slugs.
>
> There were things under things, as well as things inside things.[24]

The constellation of things under and inside things is spatial—sliced runnels, little pockets or open graves, burial-chambers—as well as insectile: H.D. develops an understanding of the unconscious as larval, frantically circled by unknown, unborn things, waiting to hatch. Here is the enigma of the (racialised, orientalised) unconscious, entered through the 'sanctum' of Freud's consultation space, unfolding the strange, spellbinding spatialities and temporalities of the unconscious and its frantic, obscure dynamics.[25]

The relation of the third kind, as Blanchot defines it, shares its movements with the psychoanalytic discourse, then, not least because its processes are mapped as a logic of spatialised, corridic unravelling:

> a mobile-immobile relation, untold and without number, not indeterminate but indetermining, always in displacement, being without a place, and such that it seems to draw-repel any 'I' into leaving its side or its role—which, nonetheless, the 'I' must maintain, having become nomadic and anonymous in an abyssal space of resonance and condensation.[26]

The I in this relation is beside itself, simultaneously drawn into 'being without a place' and repelled from there in a double, 'internally divided' movement that characterises fascination, too.[27] This I, no longer self-subject, is nonetheless to keep making itself available—it retains the ability to deny consent—to enter and uphold such relations without 'filiation,' without form or identity.[28] This relation of the third kind refuses form: the form of the subject, the form of the other as object to be taken hold of or recognised as subject just like me, the reification of the other and the domestication and/or neutralisation of strangeness. Strangeness 'founds' this third relation, 'always in displacement,' so without foundation altogether. Between the anonymous, amorphous I and the other, strangeness is the 'interruption,' a cut that 'measures the very extent of the outside'.[29] Blanchot stresses that this is not a relation of power, noting that the strangeness of the other does not occur within my narrow horizon—established through the parameters of my knowledge or, as it may be, my ignorance—but is absolutely external to it and therefore can't be integrated and thus fundamentally interrupts being.[30] As a result, when that other, without horizon, speaks to me, or when I call upon the other in this relation of the third kind, the speech that takes place is one that is marked by that interruption, understood as contestation, an opening up; occurring in an abyssal space, it is 'unfettered to any native soil'.[31] It brings into play the enigma of the other, the outside, the neutral, the phenomenon of infinite distance that I, a 'nonpersonal punctuality oscillating between no one and someone,' have to respond to.[32] Receiving the speech of the outside means to write, to lay myself open to relations of the third kind and the experience of language: relation without relation, in a realm in which fascination reigns.[33]

Relations of the third kind are sensed, then, and can be formed, in the space of writing, also the space of fascination and of the analytic setting, where the 'very nature of being in relation' is altered.[34] Blanchot infers, rather than directly invokes, Lacan: the I as 'non-personal punctuality' resonates, for one, with the Lacanian reading of the subject as stain, caught in the other's gaze. The description of a relation without identity or 'native soil' further gestures to the rift or cut that, according to Lacan, divides the so-called human subject from itself. Reading Blanchot while 'in love' with Lacan (and vice versa) indeed yields an echo-land if we contest André Lacaux's claim that any approach of Blanchot through Lacan with love [*aimer Blanchot 'avec' Lacan*] is a product of 'thirty or twenty years ago'.[35] This, at any rate, is a rather odd statement to make, considering the importance of transference in treatment, which begins, after all, with love. More than that, though, this relation of the third kind is, beyond any doubt, a 'side-issue' of, or akin to, the opaque processes of psychoanalysis, including the phenomena of transference and of fascination, placed into alignment. In this relation, a relation and space of dispossession, fascination makes the I lose its bearings and 'except[s]' it 'from the problematic of being' in the sense that the fantasy of the phallic and so-called human I is 'disarmed,' turned aside.[36] What is at stake is a *Verwandlung*, from the I into no one and, as we shall see, into the 'residue' of the subject as insectile subjectile.

If relations of the third kind constitute a psychoanalytic scene—without folding the subject back into the status quo, which, for much of the French Left, summed up the state and politics of psychoanalysis in France in the '50s[37]—and if these relations correspond to an event of fascination, then it follows that another element presses its claim on our attention. The element that cuts in, here, is the insect, associated with both the phenomenon of fascination and the concept of metamorphosis. Roger Caillois, discussing mimicking insects, argues that the decisive factor in, or rather before, a creature's assimilation to space is fascination and that insects as objects of study and collection similarly appear as fascinating.[38] They also function as traces beyond the so-called human, an indication of the slippage of that discourse of the 'human' happening within the subject itself. Insectile subjectile, accordingly, documents that slippage or glitch of the insectile other inscribed but usually unacknowledged in the subject. In *A Thousand Plateaus* (1980), Deleuze and Guattari speak of the dismantling of the 'organism' of the subject by nonetheless retaining what they call a 'minimal subject':

> You have to keep enough of the organism for it to reform each dawn; and you have to keep small supplies of signifiance and subjectification, if only to turn them against their own systems when the circumstances demand it, when things, persons, even situations force you to; and you have to keep small rations of subjectivity in sufficient quantity to enable you to respond to the dominant reality. Mimic the strata. You don't reach the BwO [Body without Organs], and its plane of consistency, by wildly destratifying.[39]

The BwO is 'what remains when you take everything away,' that is, the fantasies and regimes of signs that determine subjectification, including the fantasy of the so-called human.[40] The minimal subject mimics the strata, what Deleuze and Guattari elsewhere call acts or organisations of capture;[41] in other words, it pretends to have form when it is actually all becoming. This process of becoming is not limited to a becoming-insect but allows for all sorts of temporary assemblages; my interest, here, concerns itself only with the insectile subjectile as a manifestation of that minimal subject articulated as insectile.

Entomological Fascination

The enigma of fascination is central to subject formation as Lacan argues in *Seminar II* (1954–1955): the subject-to-be, passing through the mirror stage, 'hangs completely on the unity of another ego,' whose perfect mechanism fascinates.[42] Yet fascination concurrently has the potential to rend the I apart and is bound up with a dreamy terror, being rooted to the spot. The ambiguity of the state of fascination is captured in that act of stiffening, as Freud shows in 'Medusa's Head' (1922). The I erects itself, demonstrating that it remains in the possession of the phallus to ward off the threat of the (monstrous female) other which petrifies, lures the subject into a 'sunken place'.[43] Blanchot, though, as we have already seen, interprets fascination and the fascinating space of writing in terms of a potentiality: to be able to '[return] to myself in the form of Someone,' that is, to no one 'in particular,' a 'faceless third person'.[44] The subject in this case is not something that is constituted by this event, in this space or dead time, but is drawn outside, as an 'impersonal being,' away from the world.[45] This pure exteriority, as Foucault explains, is a 'form of thought [...] that stands outside subjectivity,' where the I is no longer bound by a discourse that holds it in place, in proper order, as 'responsible agent' and representative of the discourse of a phallic subjectivity.[46]

The framework or environment that I would like to establish for Nabokov's *Ada or Ardor*, the first text to be considered in terms of its engagement with the potential of an insectile subjectile, is that of fascination, of the novel operating in relation to a space and time of being fascinated. I don't claim that *Ada* is faithful to this thought from the outside—far from it, considering its focus and *raison d'être*, an old man dreaming of rapacious youth, as Michael Wood thinks—but that the novel is assembled (because there is mastery here) in response to a milieu of fascination.[47] I'm well aware of the apparition of the author as a 'screen for our desire,' as Oliver Harris writes in relation to William Burroughs, of the alluring figure of the writer 'behind' the work—I admit being prone to hallucinating Thomas Pynchon as just such a figure—but the orchestration and motivation of the narrative in *Ada* remain too obvious here, the figure of the author glimpsed a product of deliberate manipulation.[48] Consider the obvious pleasure Nabokov must have gotten from integrating his lepidopteran knowledge into the structure of the novel, as Liana Ashenden remarks;[49] the many parallels between Van and Nabokov, or also Ada and Nabokov, or between *Ada* and *Lolita*; the centripetal arrangement by which different elements of Nabokov's life cohere at Ardis, where most of the memoir is set;[50] the inability to notice suffering in Van's narrative, making it appear callous, a shortcoming that, Wood suspects, affects Nabokov, too, 'not [seeing] as much as he might'[51] concerning those that are pushed aside to enable the possibility of a dyadic happiness. The definitive image that persists of the memoir/novel and love affair is that of Ardis as a snow globe, an image provided by Lucette, brutally sacrificed to Van and Ada's love affair. Closed in on itself, in a sphere of 'solid crystal with snow falling as if forever all around,'[52] the snow globe on the one hand expresses Van's desire to arrest and preserve time, thereby articulating his desire for fascination itself: 'the desire for fascination,' writes Steven Connor, 'is the desire for arrest,' for a state in which the subject is at once 'enthralled and aroused'.[53] On the other, the globe also illustrates the 'form' of a thought that is locked in, static, attentive only to the frozen and precarious existence within, rendering everything else, beyond the horizon, mute and invisible, as well as expendable, objects to be used or discarded at will.

It is not, then, that *Ada*'s scene of fascination abandons the subject to an outside. On the contrary, it is an expression of total self-involvement and solipsism, of narcissistic engulfment. The world around Van is there for his consumption or enjoyment, even if such vampirism appears as a compulsion to shore up happiness. He longs to find 'Verglas,' signs which warn of slippery roads ahead but which Van interprets as 'some magical town, always around the corner, at the bottom of every snowy slope, never seen, but biding its time'.[54] The ruin of an order can, of course, always take place—that is partly what *Ada* is about—destabilised by its own logic, but what interests me above all is how *Ada* manifests itself, despite its snow globe/Verglas interiority, as a work profoundly engaging with, and the result of, fascination, whose enigma is omnisciently (omni-incestly and omni-insectly) present.[55] If writing is a scene of fascination, that scene, inward-turning as it is in *Ada*, does not, in this case, correspond to the potential Blanchot ascribes to it. Nonetheless, the novel allows an investigation into the correlations between insects and the phenomenon of fascination, in turn shaping what I'm calling an entomological fascination at the level of language.

In Nabokov's novel, Lucette tells her half-brother Van, with whom she is desperately in love, of a series of sexual encounters—'practically every night' during a particular summer—between her and her half-sister Ada, Van's sister and lover:

> So the day passed, and then the star rose, and tremendous moths walked on all sixes up the window panes, and we tangled until we fell asleep. And that's when I learnt—concluded Lucette, closing her

eyes and making Van squirm by reproducing with diabolical accuracy Ada's demure little whimper of ultimate bliss.[56]

As Liana Ashenden has argued, insects interweave with scenes of fucking and 'delicious fun' involving Ada and Van—their love story is at the heart of Nabokov's book, written largely from Van's perspective—as well as between Lucette and Ada: the insects' appearance is intimately, 'incandescently,' linked to incestuous sexual desire in the novel.[57] There are many reasons for this correlation, not least considering Ada's '*raffolement*'—a neologism I derive from the verb *raffoler* used in the novel, meaning to be passionately seized—with orchids and insects, specifically but not exclusively butterflies. Both form objects of fascination for Ada that Van deploys for their sexual connotations in his memoir of their lifelong affair.[58] Ashenden writes that Van is not interested in these objects per se and is concerned with them only in so far as they bear onto his sexual relationship with Ada. Yet it's not easy to disentangle the elements of *raffolement* or fascination—however much these two phenomena might differ—in the narrative, to determine what infatuations 'belong' to whom or from whence exactly they emanate. The very nature of such subject–object relations is, as Oliver Harris has shown, one of 'excessive identifications' that surpass individual subjects.[59]

In his book *Williams Burroughs and the Secret of Fascination* (2003), Harris identifies Burroughs as a 'point of excess, a kind of inner extremity' functioning across the latter part of the 20th century in the United States, compelling both admiration and repulsion.[60] He argues that 'at the heart of fascination, there is no thing, no single and static material object, but a complex and mobile relation' that paralyses the mesmerized subject, held captive in a curious dimension.[61] There are significant discrepancies in a subject's responses to either being fascinated—to which there is, as mentioned earlier, a terrifying dreaminess—or the adoration of *raffolement*. *Raffoler*, incorporating the word *folie* [madness], suggests an instinctive and turbulent physical relation to an object, whereas fascination is ambiguous, paralysing the subject in the object's gaze.[62] The ambivalence that characterises fascination is absent from *raffolement*—in the latter state, the subject does not seek to recuperate the control it has lost, and it enjoys yielding itself up to the object—but, even so, an affective alignment emerges precisely in the excesses compelling the subject beyond itself. In fact, the erotic entomological patterns that Ashenden traces in her article revolve around, and keep returning to, fascination, a phenomenon involving Ada and Van but also Nabokov as well as whomever reads the novel in complicated circuits of desire generated in and outwith the narrative. What all this means is that to attribute a fascination with insects to Ada but not to Van—who, for example, associates 'fascinating' fireflies with the 'harassments of sweat and sperm,' that is, nocturnal masturbatory 'ordeals' at the start of his early love for Ada—is to be blind to those circuits that fundamentally constitute and determine the attachments forming (in) *Ada*.[63] In other words, entomological fascination, while predominantly traced to Ada (or is it Nabokov) in the novel/memoir, can't be confined to her alone; it describes a number of relations, or regions of intensities, between subjects and objects, none of which remains unaffected by its energies.

By way of an example, I propose looking at Van's description of Ada at Ardis Park while sunbathing, taking on attributes that cast her as insectile. Her 'long straight hair' appears 'bluish-black in the shade,' revealing its vibrant colours under a 'gem-like sun,' its 'texture, gloss, and odour' marking the beginning of Van's *raffolement* with her, which remains constant up until the point of his death. His descriptions of her hair assemble a set of characteristics—an iridescence emerging in particular conditions of light; a texture and gloss reflecting, and reflective of, a gem, the glaze itself, as Freud has shown, suggestive of the fetish—often associated with a *scarabaeus* or scarab beetle, feeding on dung but resembling a jewel. (The beetle, 'hard and glossy,' is the subject of Edgar Allan Poe's 'The Gold Bug' (1843); in the story, the bug's remarkable weight indicates that it consists, in fact, of pure gold, further pointing to buried treasure that, once unearthed, flashes up its riches in a 'glow and glare'.[64]) If Ada's hair, with its colour-shifting properties and optical mechanisms resembling the *scarabeidae* species,[65] is fetish object, tickling Van's legs, creeping into his crotch, entwining and coiling around his rising 'column,'[66] the eroticism at work evokes Ada as creature profoundly shaped by an entomological fascination that might have 'originated' with her but similarly acts on Van. Scarab beetles' signalling to conspecifics through the exocuticle reflection of polarised light corresponds to Ada's means of communication, composing a 'map,' in Deleuze and Guattari's vocabulary, between Ada and *scarabeidae*.[67] The entire narrative of the memoir is generated by this particularly entomological fascination, whose enigmatic operations involve multiple, or iridescently layered, subject–object relations.

I'm interested in fascination and iridescence, if you wish, the latter less an optical than a metaphorical phenomenon, in the sense of being attentive to a diffusion and scattering, a type of crowding and layering or, to put it in Van's words, a 'group of ghost cells' clustering in and around concepts and affecting, more generally, the signs of a language.[68] Van mentions ghost cells, the 'possibility of some prolonged form of disorganised consciousness' existing beyond death, when visiting Philip Rack, Ada's critically ill music teacher and onetime lover, in hospital, and unkindly imagining an 'eternal Rack,' an 'infinite Rackness' persisting after the teacher's passing. This 'infinite Rackness' consists of 'tiny clusters of particles still retaining Rack's personality, gathering here and there, clinging to each other, somehow, somewhere,' his German accent 'probably constitut[ing] his most durable group of ghost cells'.[69] These ghost cells suggest swarming and excess, supplements always already 'hollowing out' presence.[70] *Ada* demonstrates that the possibilities and slippages of meaning, the logic of supplementarity, are all linked to the insectile, to that entomological fascination that determines the memoir which, right from the start, plays (excessively, obscenely) with 'contortions of tongue or syntax,' as Wood puts it.[71] Tiresome puns chase the novel onwards, whose title is indicative of the 'ghost cells' haunting Ada, her very name clustering impressions and references, 'shadow words,' recalling Blanchot's term. The other central axis, mentioned earlier, is the association between insect and incest, but the entomology of language, proceeding from these central axes—it really is just one, a 'point of excess' and 'inner extremity' binding together Ada, insects and incest—is articulated and proliferates in a number of ways, involving both the liveliness of language and its stoppage:

- Ada's 'spectacular handling of subordinate clauses,' as if themselves a kind of larvarium: clauses behave like larvae, passing through a succession of stages and formulations; her frequent asides, where she adds information to Van's narrative, often concerning insects, in parenthetical enclaves;[72]
- the word games, where 'insect,' travelling via 'nicest,' ends up, of course, as 'incest'; Van's 'reviewing' of dreams, including 'professional dreams,' dealing with his work as a writer, in which 'somehow, [...] the book had already come out, [...] with a typo on every page, such as the snide "bitterly" instead of "butterfly" and the meaningless "nuclear" instead of "unclear"'; the cryptograms the lovers send to each other, compositions of 'inserted afterthoughts, deleted phrases, rephrased insertions and reinstated deletions with misspellings and miscodings' or the riddles they whisper in each other's ears linked, in one instance, to a Silesian river ant;[73]
- the deep entanglements of entomology and simile, metaphor and metamorphosis, like the instance of Van remembering 'lovely, naked, shiny, gaudily spotted and streaked sharkmoth caterpillars' and a 'little Vapourer fellow,' whose 'black coat [was] enlivened all along the back with painted tufts [...] of unequal length, like those of a fancy toothbrush treated with certified colours.' He continues: 'that kind of simile, with those special trimmings, reminds me today of the entomological entries in Ada's diary,'[74] an observation followed by a dash, signalling an interruption. The whole page is, in fact, full of interruptions and bracketed asides, exclamation points, commas and semi-colons, 'friendly spirits,' according to Theodor Adorno, or also to be interpreted in a 'long tradition' connecting the bodies of flies or other insects to punctuation marks.[75] Through punctuation marks, we slip back into simile and metaphor, the larvarium of syntax. One of Ada's favourite words in the English Language—she is fluent in a few—is the word 'husked,' because it '[stands] for opposite things, covered and

> uncovered, tightly husked but easily husked, meaning they peel off quite easily'.[76] Clothing is a husk to be peeled off in this eroticised scene with Van which proceeds from recalling that game of anagrams ('insect; nicest; incest')—in German, to dress up means to 'throw yourself into the husk' [*sich in die Schale werfen*]—and so is language, a husked hollowness that, at its missing core, keeps moulting into new 'forms'.

The examples above, by no means exhaustive, roughly fall into three categories, the first of which we might persist in calling the larvarium of syntax, while the second refers to the Freudian scene of writing, the impressions on the 'text' of the psyche left by the fascinating object, that is, Ada and insects—Ada-insect, a type of Brundle-fly—totally shaping said psychic text.[77] The games, the cryptograms, the dreams, the 'oneiric word-plays,'[78] the third aspect of metaphor and simile, the 'interior' of the memoir, these are all effectively informed by and release both voluntary and involuntary traces of Ada and Van's relationship, whose poetics (and erotics) is that of the insectile, itself indicative of that which breaks or cuts in as well as of that which transforms. The locus of fascination—bearing in mind that fascination has no concrete locus—involves, as such, also language, performing its excesses and nestings in Nabokov's novel, so intent on taking pleasure in a writing that writhes.[79]

Swarm-Being, Language-Nests

According to Derrida, for whom the figure of a single ant becomes the vehicle to let himself be swept along by associations, the result of a type of automatic writing, the experience of language is akin to a *fourmillement*: to give yourself in to the word 'ant,' always synecdoche for an innumerable multiplicity.[80] A meditation (though that sounds too measured) on sexual difference, 'Fourmis'—further reflecting on dreams, the economy of the gift which exceeds the 'mercenary and mercantile circle of salary'—is also about the pleasure of the text or, rather, performs the text as 'site of bliss'.[81] Roland Barthes proposes that a text of bliss, always bound up with excess and loss, arises out of a cut, or bears the 'trace of a cut,' disturbing or destroying the consistency of the subject.[82] 'Fourmis,' so attentive to the function of the cut—etymologically, 'insect' comes from cutting—is an open body, constituted through and inviting yet more incisions. Derrida likens deconstruction to a parasitic reading or interpretative practice, ceaselessly cutting in, putting words in quotation marks to lift them out of their given context, proliferating the meaning of terms. This approach generates acts of writing, too, parasitic texts that are as much an assemblage as a dismantling. Reading 'Fourmis,' or even merely glancing at it, skipping over it, reveals it as a gush of words, a swarming: it is a text marked by the insectile, and its constant ruptures serve to expound the murmur and interruptions of language.

I offer an aggregation, at this point, to develop the proposition that the otherness that haunts writing or language is insectile, even if that comes at the risk of pushing it into some kind of form. As we have seen, the insectile can be both total form (the armoured subject or text)[83] and formless, a 'figure' for the absence of form. Context assembles an armature of sorts; an act of interpretation, if it is to appear as good form, that is, convincing and able to provide supportive evidence, is dependent on two things, as Colin Davis reminds us in *Critical Excess* (2010). These two things are context and coherence, supposed to regulate interpretations, yet both these aspects are, at least to a certain degree, either assembled (context) or forestalled (coherence). The armature of the context, here, is that of the insectile, provided by Nabokov's *Ada or Ardor* on the one end and Lispector's *The Passion According to G.H* on the other, the latter exalting in the loss of form, 'the lost system of good taste'.[84] In the space between these two texts, both of which deal with the phenomenon of fascination in conjunction with insects, operates a little text-machine, focussing on the processes of language—including punctuation marks; Freud's concept of parapraxes [*Fehlleistungen*]; delirious (over)readings—as swarm-being or language-nest. The argument, as such, seeks to demonstrate writing as movement of cutting, of cutting in from elsewhere, of otherness always already nesting within. These 'spores' of the insectile generate relations that turn the subject away from the fantasy of the so-called human, the form of the I as subject of command and control.

Speaking at a conference about Beckett and animality in 2009, Steven Connor remarks that there is a 'strong affinity between swarming and the jostling of words,' while further drawing attention to the 'dipterous contour' of the comma, deriving

> from the Greek *kopma*, from *koptein*, to strike or cut, the mark of elementary division, dividing off the smallest unit of grammatical sense, that is nevertheless itself not quite entire, that, like the insect has duality or division within it, between body and tail [...].[85]

Connor talks about the comma in *The Unnameable* (1953), linking it to the function of the fly, not least because it agitates textuality, the comma's interstitial appearance itself indicative of the intrusions of a fly. He cites Ambrose Bierce, who suggests that punctuation marks stem from fly-specks and proposes that they exist as a 'kind of noise,' neither figure nor ground, of any given text.[86] Connor's paper demonstrates the ways in which the insectile more broadly determines the inscription of language (written or spoken), an 'event' or condition of existence not only evident in 'Fourmis' or across Beckett's work. In Jonathan Lethem's *Motherless Brooklyn*, the central figure, Lionel Essrog, suffers from Tourette's, forced to expel words that itch in his brain:

> Lionel, my name. Frank and the Minna Men pronounced it to rhyme with *vinyl*. Lionel Essrog. *Line-all*.
>
> Liable Guesscog.
> Final Escrow.
> Ironic Pissclam.
> And so on.[87]

The debilitating disorder that is Tourette's—a subject for Beckett, too—constitutes one of the most severe and intolerable manifestations affecting the mechanisms of language, convulsing, in the passage above, into phonetically related expressions of Essrog's name, whose buzz reverberates throughout its various iterations: Essrog, Guesscog, Escrow, Pissclam. The figure of a name collapses into ground or noise, the autonomous body asserting itself beyond the will of the I, disturbed in its onward or forward motion and stuck instead on a terrible compulsion to repeat.[88] Tourette's is an impersonal phenomenon, in that it is the materiality of language, the 'reptilian brain' or rather the insectile, that articulates itself beyond the intentional subject. In his article on the 'poetics' of Tourette's syndrome—the word poetics itself is tantalising, a 'tic by definition'[89]—Ronald Schleifer argues that what links poetry and Tourette's is an engagement (totally involuntary in the latter case) with the 'resources' of language, its material and frequently repressed mechanisms.[90] He writes that the 'strange fascination' of Tourette's has to do with the force of its discourse, both verbal and motor (or verbal as motor), but that poetry or the literary more broadly shares in this fascination, the rhymes and repetitions, the sounds and rhythms of language.[91] Schleifer is arguing not that writing is an 'impersonal medical condition' but that it calls forth impersonal states, states of fascination to refer back to Blanchot, in which the I is held in a realm of shadows, the *fourmillement* of words.[92]

Without coming close to the social consequences that the syndrome has, or the enormous personal efforts demanded in order to keep the 'echo-chamber skull' tic-free for even a few moments,[93] Essrog's verbal ticcing—a state of practically constant interruption; the insectile as the narrator's fundamentally experienced condition—resonates with lesser phenomena, phantoms emerging from the unconscious that Freud describes in *Psychopathology of Everyday Life* (1901). The psychic disturbances Freud concentrates on are, to begin with, the forgetting or displacements of proper names, which 'keep returning and force themselves on us with great persistence,' often appearing in divided forms, one pair of syllables recurring without alteration, the other containing 'a numerous and miscellaneous set of relations' to a repressed topic driving the process of displacement.[94] *Psychopathology of Everyday Life* consists of an archive of distorted matter, liquifying words, thought-content deviated by trauma; the list of examples for each of the interferences is exhaustive, demonstrating 'fluctuations in the control' over language.[95]

In the section about the forgetting of foreign words, Freud is drawn to metaphors of water, of things gliding out of grasp, prompted by the first instance of this type of interruption he provides. Scanning the pages, looking for metaphors of insects, I misread the word 'misquoted,' which shifts into 'mosquitoed,' a mistake evidently brought on by wishful thinking, the (I fear) convulsive attempts to fit my research focus into Freud's text. My 'preparedness' for finding the insectile everywhere alters the material I'm looking at, especially if distracted: *Calliphora vomitoria*, blue bottle flies, keep entering through my open window, their buzz loud enough to break my concentration. In addition to wish fulfilment, the possibility of insects appearing in the field of vision, literally or metaphorically, peripherally and centrally, concerns anxieties having to do with overreading (whether or not that exists), being a bad reader and, beyond that, being unfit for my profession.[96]

The preconditions for a misreading such as the one above (misquote = mosquito) arise in the (paranoid) function or quest of research, self-doubt, as well as similarities in the 'verbal image,'[97] none of which manages to dispel the insectile, neither from Freud's study nor from the *fourmillement* of language more generally. The insectile is there, to recall, in the etymological kinship between insects and cuts. Freud talks about words divided, interruptions, irruptions, insertions, incubations, interferences, disarticulations and contagions, aspects that, as we have seen, are closely associated with, and often visualised as, insects, making the kinship metaphorical too. In *Angels and Insects*, A.S. Byatt writes that metaphors are a kind of metamorphosis, carrying one idea over into another. In 'Morpho Eugenia,' one of the two novellas that constitute *Angels and Insects*, the word 'insect' repeatedly transfigures, from 'living jewels' to the vehicle, in a game of Anagrams, to confront William Adamson with his wife's incestuous relationship:

> At one point, finding himself with PHXNITCSE he suddenly woke up, and found himself able to present Matty Crompton with INSECT even though that left him with an X with a demon on it. Miss Crompton, her face heavily shadowed in the lamplight, gave a small snort of laughter at this word, considered it for some time, rearranged the cards, and pushed it back to him. He was about to point out that the rules did not allow of returning the same word with adding or subtracting a letter, when he saw what she had sent him. There it was, lying innocently in his hand. INCEST.[98]

The game reads like a Freudian case study, while referencing Nabokov's *Ada or Ardor* (as well as *Finnegans Wake* (1939)): Adamson, concerned with the letter X and being left to hold a 'demon,' unconsciously presents 'insect' as veil or screen to hide the knowledge of his wife's incestuous relationship with her brother. He effectively (dis)places his trauma (the primal scene) into the hands of another, Matty Crompton, who drags it out into the open and pushes it back to Adamson, at first sight unable to recognise his friend's intervention. Incest remains, briefly, under the cover of the carapace 'insect,' at once Adamson's defence to ward off the traumatic scene and the means by which the repressed returns. This recurrent sense of transfigurations—whatever the motives, conscious or unconscious—is apparent throughout the novella, so alert to the ways in which insects are made to figure.

If misreadings and other parapraxes have to do with the subject's lack of command, disturbances that reveal the slippage of the I, the context to these 'errors' is a distortion of some kind, according to Freud, elements that, unforeseen, cut in, disturb, and insinuate themselves. There is, though, also the notion of excess, itself linked to overreading, an 'exorbitant' practice that transforms a text (or, depending on perspective, mutilates it and thereby prevents a 'good' interpretation).[99] In many ways, 'Fourmis' exemplifies this state of being disturbed and led 'astray,' an urge to the truth, perhaps, or the refusal of the good form, what Hans-Georg Gadamer calls the '*Vorgriff zur Vollkommenheit*,' the anticipation of perfection, the perfect unity of sense.[100] Once you begin looking up etymologies of words, you might find yourself, like Derrida or Matty Crompton—thinking that insects are 'walking figures of speech,' morphing from one form into another—'carried away' by their implications, their potential for metamorphosis.[101] The novella 'Morpho Eugenia'—named after a neotropical butterfly native to French Guiana (the history of colonisation casts its shadow on the English manor,[102] according to Michelle Weinroth) and Adamson's incestuous wife—evolves the 'form' of the insectile through a series of stages. The story begins with a ballroom dance, 'shimmering girls' in 'shell-pink and sky-blue, silver and citron, gauze and tulle' winding past Adamson, a Victorian naturalist, whose collection of insects was lost in a shipwreck.[103] The attention to matter is telling, the girls moving into a field of visibility defined by a natural historian's gaze, trained to observe and mesmerized by surfaces, forms, colours, substances. This gaze operates throughout the novella, alighting on the dress of English men, wearing 'carapaces like black beetles,' or young women in 'close-fitting, unornamented bodices'—themselves a type of carapace—entering rooms 'like a cloud of young wasps' or, at other times, 'standing still, cocooned in silk'.[104] Then there are the analogies between insects and other unwanted creatures, unusually characterised by kindness, even despair against the exterminating or immiserating forces of racist-capitalist subjection. Adamson, encountering Mrs Larkins, a kitchen maid catching and burning 'seething [masses] of black beetles,' aligns her with the 'imprisoned Coleoptera, struggling and hopeless'.[105] The correspondence between the insectile and the preterite or subaltern is culturally prevalent; this logic of racialisation informs, for example, the narrator of Lispector's *The Passion According to G.H.*, who likens her former maid to an 'African queen'. Further descriptions of the maid in *Passion* render her as insect specimen, her body like a shell, thin, hard and smooth, with a 'black and motionless face' and 'wholly opaque skin,' whose 'closed-off blackness' gives the impression of being 'only [...] an external form'.[106]

Metaphors migrate ideas, nest them into other forms; that's their essence or, I should say, their rhythm, their vector of becoming or, conversely, as so often in the case of racialisation and abjection, their force of hardening, of arrest, capture and annihilation. The movements and figurations of the insectile, the correspondences between language and insectility, do not end there, though the ideological constructions of self and other certainly are the most pernicious and require the most persistent dismantling. Nor is the swarm-being of language limited to a comma's disarticulations or the glitches that enter, unforeseen, into communications. In Donna Tartt's *The Goldfinch* (2013), learning French is equated to insect collecting—'too many vocabulary words, [...] *Nom and Prénom*, species and Phylum. It's only a form of insect collecting'—whereas Nabokov talks about the process of translation as entomological transposition in his memoir:

> For the present, final, edition of *Speak, Memory* I have not only introduced basic changes and copious additions into the initial English text, but have availed myself of the corrections I made while turning it into Russian. This re-Englishing of a Russian re-version of what had been an English re-telling of Russian memories in the first place, proved to be a diabolical task, but some consolation was given me by the thought that such multiple metamorphosis, familiar to butterflies, had not been tried by any human before.[107]

In Tommy Orange's *There There* (2018), Tony Loneman, a 21-year-old born with foetal alcohol syndrome he calls 'the Drome,' painstakingly learns to read at night, but the letters 'move on [him] sometimes, like bugs'. Even when they're still, he has to 'wait to be sure they're not gonna move, so it ends up taking longer for [him] to read them than the ones [he] can put back after they scramble'.[108] On the page, words leave 'trace marks' as if 'made by some kind of wriggly insect,'[109] as Joyce Carol Oates writes in *The Sacrifice* (2015); in both Oates and Orange's works, language squiggling like bugs is a phenomenon associated with the subaltern. The inability to 'master' language—the way it deteriorates, leaving no firm foundations despite its textuality—forms part of policies of racialisation like lack of access to education, even if these aspects (deterioration, faltering, etc.) also and more broadly dissolve the notion of an intentional subjectivity.

As always, there is a counterpoint to the insectile *informe*—as movement of loss, displacement, constant transformation—when it comes to language, involving communication techniques inspired by bees or ants as technologies of form: on the one hand, the 'Drome,' if you allow a repurposing of Loneman's condition to describe the stuttering and murmuring of even normative language use, in subjects that are not traumatised, affected by debilitating symptoms or racist politics and, on the other hand, efforts akin to a *Vorgriff zur Vollkommenheit*, a form of communication so disciplined and pared down to eradicate misunderstandings. Charlotte Sleigh discusses such initiatives in her work and mentions, for example, the British linguist Charles Kay Ogden's development of Basic English, proposed in 1930. There are, as she shows, other international languages modelled on the communicative economy found in insect colonies, which later, most notably at the Macy Conferences (1941–1960), serve to inform cybernetics, modelled on understanding the 'intelligence' of non-human actors in networks of ecological relations. These proposals proceed on the basis of swarms as systems, which are thereby bestowed with form, able to send unambiguous signals back and forth, clearing away the noise that distinguishes communication between so-called humans. Sleigh cites the myrmecologist William Morton Wheeler, who remarks that

> perhaps, one's attitude towards words should be that of the observer in the tropics toward insect fauna. Some words are like gorgeous butterflies and harmless, others [...] like blood-sucking Diptera and Hemiptera which are vectors of subtle viruses. Are there not also parasitic [,] symbiotic and predatory words like the corresponding groups of insects [?] Also mimetic, warningly and protectively coloured words? The entomologist resembles [...] the philologist. Like insects [,] words lack meaning except as their behaviour in connection with their [...] environment is brought into the picture to form a context.[110]

This is an extraordinary passage, not least in its metaphorical (yet again orientalising) approach to language, beautiful form, harmful virus, dissimulating husk, gathering its meaning only in context to its *Umwelt*. Beyond the metaphorical, though, language functions as a type of insect technics, to borrow Jussi Parikka's term—evident already in the discussion above—in that it involves embodied processes of becoming, a conception of words as exoskeletal, all outside with no insides, as a vector of a viral, swarming thought. Wheeler emphasises its morphological aspects with the objective to anticipate, preserve or impose form, to purge language of its 'improprieties' and parasites, thereby at once indexing the insectile to form and the *informe*, the latter to be expelled by an 'order of insects,' a dream of shelled perfection and hard defence. This discussion about the relation between language and control, or language and harm, becomes even more urgent in a context different from Wheeler's (he died in the US in 1937). In 1947, Victor Klemperer published *LTI*, about the language—'Lingua Tertii Imperii'—of the Third Reich. Klemperer's book derives from diaries kept between 1933 and 1945 and is composed of a series of vignettes analysing Nazi order-words like *Das Volk* or *Der Jude* but also less evident expressions like *aufziehen* [winding up] which, like *fanatisch* [fanatical], undergoes a change of value in Nazi Germany. Usually associated with mechanical activities and used dismissively, aufziehen is deployed in relation to feats to be proud of, to raise something up to higher spheres, while *fanatisch* becomes, as it were, a cardinal virtue, denoting unwavering, thanatoid devotion. Klemperer's study time and again refers to LTI as *Gift* [poison], itself a Nazi trope, that totally shapes the processes of speech and thinking, conscious and unconscious, even of those who opposed Nazism and/or were its victims, and frequently despite every possible defence (like a forensic or entomological eye) mobilised against it. His book is alert to how LTI insinuates itself, spreading across the entire German language, in which all articulations must now be public, addressed to the *Volk* and thereby be either invocation or incitement to fanatical racist and antisemitic violence. The aim of the diaries, beyond their enormous psychological necessity for the persecuted Klemperer, is the desire to resist the unconscious sedimentation and subjectivation of *Nazideutsch*. The philologist-entomologist or thinking subject's duty is thus not only to observe but to intervene in the 'formatting' of language, the principles of order according to which it takes shape but which are not eternal, despite the efforts to insist on a regime's perpetuity, which is not to say that its order-words have disappeared. To finish this aside: the point is that language as 'insect fauna' is a matter of formatting or formalisation, but these processes of form and forming are always prone, as we have seen, to flip into their apparent opposites.

In 'He Stuttered' (1994), an essay on a poetics (tics, tics, tics) of a disintegrating language, Deleuze speaks of stuttering, murmuring or stammering as an 'affect of language' and a creative practice that 'makes language grow from the middle, like grass; it is what makes language a rhizome instead of a tree, what puts language in perpetual disequilibrium'.[111] He mentions Kafka, Balzac, Beckett, and Melville, along with a few others; in the last part of this chapter, I supplement this list to consider Clarice Lispector's *The Passion According to G.H.* as a manifestation of just such a practice of rhizomatic writing, of a language in disequilibrium mirroring the relation of the third kind occurring in the novel. Blanchot, as we have seen, describes this relation as a turn toward the neutral, the most radical way in which the I is called into question. The neutral, for Blanchot, is 'an essentially errant word' because 'it is always cast out of itself' and refers to an 'infinitely distended outside,' a similarly nebulous (*informe*) term and a 'dimension' of de-structuring.[112] Effectively, the outside is an 'experience of impossibility' that manifests itself in the following traits:

- a 'present which does not pass, while being only passage';
- 'that which cannot be let go of, while offering nothing to hold onto';
- a presence so excessive that it denies access and reverses into absence, 'without leaving anything in which one might absent oneself from it'.[113]

Dead time in which the I cannot be present to itself, nor to the passing of time; a state of suspension that captures the subject but which it cannot grasp; an overwhelming and opaque other functioning as becoming, without orientation: the outside as zone of fascination. The 'form' of the subject is being drawn away from itself and from the world and can, here, meet the other without seeking to determine or appropriate it, since the I has become no one, a third person, minimal subject, unable to assume command or stake claims. This chapter ends, then, with Lispector's novel, on the one hand, because it demonstrates that fascination functions as a radical subjective undoing, prompted by a cockroach that the narrator comes across in a room in her flat, vacated after her maid quit, and on the other, because Lispector keeps evoking the neutral, linking it to the cockroach's body which, after emerging from a wardrobe, is crushed by the door the narrator slams shut. The form of attention resulting from this cut is a glide into the neutral and an exploration of the bursting (and resilience) of forms, specifically the form and fantasy of the humanised subject.

It-self

The Passion According to G.H. concerns a white, female, bourgeois sculptor, G.H., encountering a cockroach in her former maid's room and being undone by this encounter. A mirror stage of sorts is taking place here, but rather than finding the form of the subject, the scene shatters the humanized I. Seen in this light, the story is a retelling of Lacan's fable of subject formation in *Seminar X* (1962–1963), where he pictures himself, wearing an unknown mask, faced with a praying mantis, unable to see his own image 'in the enigmatic mirror of the insect's ocular globe'.[114] Lispector's novel recreates this scene of fascination with an opaque other, whose 'neutral gaze' arrests G.H., becoming remote from her properties.[115] That gaze, returned from an object, fractures the subject, 'sliding away' from itself, as Lacan puts it in *Four Fundamental Concepts* (1973): subject becomes stain.[116] The interrelation taking place between G.H and the cockroach is structured according to race (and class), considering that the roach is aligned with and functions as racialised other; the contact with said other, produced as radically different, sets the scene for G.H.'s dismantling, because experiencing herself as both other and object, perspectives that the racialising subject usually eludes.

At the breakfast table the morning of the encounter with the cockroach, G.H. sits 'framed by [her] white robe, [her] clean and well-sculpted face, and a simple body,' eating 'delicately what was [hers]'. Later, in the maid's room, this fantasy of whiteness, property and the good form falls apart:

> And I was seeing, with fascination and horror, the pieces of my rotten mummy clothes falling dry to the floor, I was watching my transformation from chrysalis into moist larva, my wings were slowly shrinking back scorched. And a belly entirely new and made for the ground, a new belly was being reborn.[117]

What happens here is a metamorphosis in reverse: the imago shatters, 'regresses' into chrysalis and back into larva, the process yielding not a form but the *informe*, a 'thing-part,' associated with the ground, in terms of what lies beyond the human, at its inhuman 'root'.[118] The narrative stutters around this scene: each new chapter repeats the last sentence of the previous section in an operation of constant returns. There is no progression, only a marooning—in that room, without an awareness of time passing—and a becoming, that is, a *Verwandlung* without term, transforming the so-called human into what Lispector calls the 'it-self'.[119] The latter un-forms the so-called human, a sculptor, no less, who likes to own and arrange things, to shape them into beautiful form.

Passion turns on questions of form and of losing form, including at the level of the narrative itself. Claire Williams notes that while the plot is minimalist—it really concerns only G.H. entering the maid's room, crushing the cockroach in the cut of the wardrobe and being transfixed by its gaze—the 'form of the narrative' circulates 'multiple historical, mystical, philosophical, and religious allusions'.[120] She describes the text as one of radical openness, because of the sheer overabundance of allusions to follow (or discard), what Deleuze and Guattari call 'lines of deterritorialisation' issuing from concepts that are not usually approached as 'intensities,' such as form or also, less surprisingly, the returned gaze, with its engrossing power to destabilise the I. Form is intensity in the sense that from the very start, in the epigraph to 'possible readers,' Lispector announces it as something to be put under enormous pressure: 'this book is like any other book. But I would be happy if it were only read by people whose souls are already formed'.[121] The address to 'possible readers' with already formed souls, bearing in mind what is to come, that is, the undoing of form, is an invitation, perhaps a dare, to submit to a text that disorganises. The epigraph thus also refers to a figure of pure potentiality which does not yet exist, which 'can be or not be': an undefined, impersonal subject.[122]

The 'incarnation' into an 'organised person' precedes the narrative which itself proceeds to disintegrate the subject, drawn from an 'I-being' into 'it-self,' a depersonalised apparition without skin, identified only through initials, G.H.[123] Skin, as we have already seen elsewhere,[124] is what gives the body form; it is the form that shapes the subject, white by default. That the encounter with the cockroach takes place in the maid's room is significant: G.H. lingers—Lispector insists on this act of lingering—at the breakfast table before 'finally' managing to get up to 'arrange' things in a space (remote, obscene; the anus, as it were, of her white apartment) she hadn't entered in six months.[125] She had expected to find a den, consistent with her racist and classist prejudice: a dark room at the '*bas-fond*' of her flat, at the end of a dark hallway.[126] What she sees instead is a vision of light, a 'calm and empty order,' a 'created void' at odds with the logic of racialisation capturing the maid, Janair with her 'black and motionless face,' as purely external form.[127] She is defined entirely by her skin—there is nothing else to see according to this 'epidermalization': the maid's racialisation affects the perception, and most likely determined its usage in advance, of the space she occupied.[128] Her room, at the end of the hallway, is the site of an unbearable encounter with the cockroach which, caught by the door, 'look[s] like a dying mulatto woman'.[129] G.H.'s initial reaction, as well as her reluctance to cross over into the maid's room in the first place, is conditioned by the racialising gaze, bestowing 'human' form only onto the subjects it always already recognises as such. In that over-determined space, which defies expectation and astonishes in its state of 'arrangement' or coherence, G.H. is drawn outside, in a movement that emanates from a returned gaze, not necessarily limited to the eye.[130] Lacan writes that the gaze fulfils the function of the stain that troubles the I from elsewhere, somewhere beside the fantasy of the incarnated, organised self.[131] Fanon, using Lacan's work, argues that the inter-racial drama between racialised and racialising subject confronts the latter with an image of 'unidentifiable' wholeness, marking the beginning of the former's function as 'phobogenic object, a stimulus to anxiety,' dislodging the racialising subject and thrusting it back into larva (i.e., unfinished form).[132] The maid's room is a crisis point in which the roach instigates profound disorder, 'inlays' the gaze of the other as that which secretly governs the subject, all the while the stain of that otherness forever escapes the grasp of the I.[133]

The mirror stage, scene of fascination, takes place as follows:

> The roach is an ugly and sparkling being. The roach is the other way around. No, no, it doesn't have a way around: it is that. Whatever is exposed in it is what I hide in me: from my outside being exposed I made my unheeded inside. It was looking at me. And it wasn't a face. It was a mask. A diver's mask. The precious gem of rusted iron. Its two eyes were alive like two ovaries. It was looking at me with the blind fertility of its gaze. It was fertilising my dead fertility.[134]

The other's gaze is inscrutable, hence the description of its face (which isn't one) as mask; its illegibility and inanimation, both racialised tropes *viz.* Sianne Ngai, locating that very otherness as the 'property' of the subject. There's more to be said about the passage above, not least concerning the reference to the gaze's fertility and G.H.'s abortion, which she is led to reflect on during her fugue state; the recurring references to rebirth, plasma, embryo, eye-ovaries and motherhood invoke the abject, that is, the absolute destruction of form. The curious sense of deep time—the roach as an 'ambassador' of an immemorial age (Lispector sounds Lovecraftian)—is expressive of time's absence, too, the dead present or 'hours of perdition' of the scene of fascination.[135] The narrative recounts what happened after the event, but to establish a clear temporal delimitation of the action, even though it happens in a single day, is difficult; the atemporality of the roach affects, as Williams shows, the chronology of the plot, always at the point of unravelling.[136] The punctuation, for one, is instructive of this near-collapse: the recounting of the day begins and ends with a number of dashes, and Lispector has, as Hélène Cixous remarks, a propensity for the colon, upon which 'everything hinges'. Cixous continues: 'If [the colon] were suppressed, one would fall into absolute rupture since what follows would not be. But if one keeps the colon, there is another precipice. There is another discontinuity'.[137] According to Adorno, considering Theodor Storm's work, dashes constitute 'mute lines into the past'; in *Passion*, though, they are indicative of suspension, the marks not of a 'burdensome heritage' but rather of its opposite: the disinheritance of the humanized form, gradually enjoyed.[138]

Those elements of suspension—the constant dashes; the colons; the ellipses trailing off, suggestive of a mind absorbed, forgetful, distracted—signal what happens in the narrative, the lingering and arrest, between form and the *informe*, so characteristic of the scene of fascination. G.H. is hanging by a thread or a dash and stands at the precipice between form, which she is reluctant to abandon, and the absence of form, a fragment and 'thing-part'. She herself describes her existence prior to the encounter as 'being in quotes,' framing herself with 'a quotation mark to [her] left and another to [her] right'. She thereby places herself, '"the I-being",' in an enclosing space that is self-curated, intentionally used to appear in certain contexts and environments: quotation marks form or engineer the 'atmosphere of the I'.[139] The maid's quarters violate that atmosphere because G.H.'s organised space and subjectivity break down: the room looks like nowhere else in the flat; it is all surface, with nothing to arrange or to push into form. Her 'location,' consequently, slips; she loses her sense of place, the 'lobster's casing' of her form.[140] Her imago shatters and she glides, as the narrative puts it, into the neutral, which Barthes calls an 'irreducible No,' the negation of form:

> Since I must save the day of tomorrow, since I must have a form because I don't feel strong enough to feel disorganised, since I inevitably must slice off the infinite monstrous meat and cut it into pieces the size of my mouth and the size of the vision of my eyes, since I'll inevitably succumb to the need for form that comes from my terror

> of being undelimited—then may I at least have the courage to let this shape form by itself like a scab that hardens by itself, like the fiery nebula that cools the earth. And may I have the courage to resist the temptation [...] to invent a form.[141]

At the beginning of her narrative, G.H. insists on form against the terror of being 'infinite monstrous meat,' which she still feels compelled to cut into pieces that can easily be apprehended by, and handled through, her sense organs. She wonders about replacing the form she lost, wishing that whatever shape arises after having shed her skin—depersonalisation occurs through the sloughing of the skin, as if in a process of moulting—it'll be a scab, if anything a solid expression of the *informe*. This work against form is reflected in the 'pure cut' of Lispector's writing, as Cixous, addressing Lispector by her first name throughout her book, notes. The '*coup du tranchant*,' which Cixous describes as a shock, a moment of surrender, 'is Clarice's practice'.[142] The cut of writing, how the story is told, ushers in a world of desubjectification as much as the story itself, in which the subject takes its place as an object among others (the function of the gaze, according to Lacan). The dimension of being that emerges through the neutral gaze of the roach is itself a 'neutral cockroach body': the self-subject, the '"I-being",' has disorganised in the 'face' of the other, renouncing the finding of a new form.

The encounter in *Passion*, with its insistence on the neutral, nonetheless retains troubling aspects, even if we approach it through Lacan and Fanon, alert to the disavowal and subsequent returns of repressed otherness so central to the construction of the subject. These aspects pertain to the roach's function as racialised or orientalised creature, the 'hieroglyphs' of its movements associated with 'the writing of the Far East'.[143] This in itself suffices to, as it were, neutralise the neutral, the many ways that the neutral, in Blanchot and Barthes' readings, refuses the hardenings of ideologies and racist order-words. Lispector's novel exemplifies the turning, the constant Möbius-strip movements, between form and the *informe*: committed to the latter, the narrative cannot totally resist the pull toward form. Possibilities of another 'order' of existence, another type of discourse, indeed exist in her writing, considering how it destabilises the structures of discourses, suspends or crushes them through sheer excess of conflicting images (the juxtaposition between dryness and the moist, for example). These certainly constitute the most celebrated elements of Lispector's work, what the scholarship tends to focus on and what my own interpretation—attuned to relations of the third kind—similarly expounds. I'm reminded of my earlier misreading, however, morphing misquote into mosquito, and wonder to what extent the desire for such relations of the third kind, for a minimal subject or insectile subjectile, influences these perspectives, to the point of ignoring or marginalising the persistence of order-words that still preserve the form of the 'proper' subject. Lispector's novel is remarkable in its explosions or interruptions of form, but to overlook the 'temptations' and persistent regimes of form in the narrative might speak to the readiness of a deconstructive duty to disarm the 'proper' subject all the while failing to fulfil that duty. 'Overreading,' notes Davis, 'entails a willingness to test or exceed the constraints which restrict the possibilities of meaning released by work'.[144] While there is no question of an overreading here (assuming that such constraints are even in place), the willingness Davis identifies with pushing a text beyond its 'limits' resonates with the urge to finally be done with the privileged subject of empire, which continues to raise its ugly head(s).[145] Its undoing—if that is the right word—comes, after all, at the expense of the racialised other, historically and epistemologically evoked as formless or else as eternally deficient form and phobic object; the fundamental operation of racialisation remains, as such, in place. It might well be that *Passion* succeeds in offering up the insectile subjectile as a figure of possibility, emerging in a text, a process of writing, as scene of fascination, but the novel, irrespective of how seductive it is with respect to the destruction of form, alerts us to the enormous power of processes of forming that undergird even those efforts to dispense with it.

Notes

1. Maurice Blanchot, *The Space of Literature*, trans. Ann Smock (Lincoln & London: University of Nebraska Press, 1989), 25.
2. Blanchot, *The Space of Literature*, 26.
3. Blanchot, *The Space of Literature*, 26–27.
4. Blanchot, *The Space of Literature*, 28.
5. Gilles Deleuze & Félix Guattari, *A Thousand Plateaus: Capitalism and Schizophrenia*, trans. Brian Massumi (London & New York: Continuum, 2004), 297 & 298.
6. Maurice Blanchot, *The Infinite Conversation*, trans. Susan Hanson (Minneapolis & London: University of Minnesota Press, 2003), 73.
7. Blanchot, *The Space of Literature*, 32.
8. Blanchot, *The Space of Literature*, 33.
9. Rosi Braidotti, 'Are Bugs to Nature as Chips to Culture? On Becoming-Insect with Deleuze|Guattari,' in *Deleuzian Events: Writing|History*, ed. Hanjo Berressem & Leyla Haferkamp (Hamburg: LIT, 2009), 146–168.
10. Susan Hanson, 'Foreword,' *The Infinite Conversation*, xxxi.
11. Deleuze & Guattari, *A Thousand Plateaus*, 263; Blanchot, *The Infinite Conversation*, 66.
12. Blanchot, *The Infinite Conversation*, 67.
13. Blanchot, *The Infinite Conversation*, 67.
14. Blanchot, *The Infinite Conversation*, 233.
15. Blanchot, *The Infinite Conversation*, 232.
16. Blanchot, *The Infinite Conversation*, 230–231 & 233.
17. Blanchot, *The Infinite Conversation*, 233. See also Sigmund Freud, 'On Beginning the Treatment,' in *The Standard Edition of the Complete Psychoanalytical Works of Sigmund Freud*, Vol. XII, trans. & ed. James Strachey & others (London: The Hogarth Press and the Institute of Psychoanalysis, 1958), 124.

18 Jane Rendell, *The Architecture of Psychoanalysis: Spaces of Transition* (London & New York: I.B. Tauris, 2017), 117.
19 Michel Foucault, 'Maurice Blanchot: The Thought from the Outside,' in *Foucault/Blanchot*, trans. Brian Massumi (New York: Zone Books, 1987), 24.
20 Rendell, *The Architecture of Psychoanalysis*, 117.
21 Sigmund Freud, 'On Beginning the Treatment,' 134. The 'side-issues' are also mentioned in Rendell, *The Architecture of Psychoanalysis*, 115.
22 Freud, 'On Beginning the Treatment,' 135.
23 H.D., *Tribute to Freud* (Manchester: Carcanet, 1985), 19–20.
24 H.D., *Tribute to Freud*, 20–21.
25 H.D., *Tribute to Freud*, 4.
26 Blanchot, *The Infinite Conversation*, 67.
27 Oliver Harris, *Williams Burroughs and the Secret of Fascination* (Carbondale & Edwardsville: Southern Illinois University Press, 2003), 17.
28 Deleuze & Guattari, *A Thousand Plateaus*, 266.
29 Blanchot, *The Infinite Conversation*, 68 & 69.
30 Blanchot, *The Infinite Conversation*, 69.
31 Foucault, 'The Thought from the Outside,' 25.
32 Blanchot, *The Infinite Conversation*, 71.
33 Blanchot, *The Infinite Conversation*, 74 & *The Space of Literature*, 31.
34 Susan Hanson, 'The Double Exigency: Naming the Possible, Responding to the Impossible,' Foreword to *The Infinite Conversation*, xxvi.
35 André Lacaux, 'Blanchot et Lacan,' *Essaim*, Vol. 1, No. 14 (2005), 41 & 42.
36 Blanchot, *The Infinite Conversation*, 10.
37 Joseph D. Kuzma, *Maurice Blanchot and Psychoanalysis* (Leiden & Boston: Brill Rodopi, 2019), 5.
38 Roger Caillois, 'Mimicry and Legendary Psychasthenia,' trans. John Shepley, *October*, Vol. 31 (Winter 1984), 16–32.
39 Deleuze & Guattari, *A Thousand Plateaus*, 178. They mention the concept of the 'minimal subject' on p. 298.
40 Deleuze & Guattari, *A Thousand Plateaus*, 168.
41 Deleuze & Guattari, *A Thousand Plateaus*, 45.
42 Jacques Lacan, *The Ego in Freud's Theory and in the Technique of Psychoanalysis: The Seminar of Jacques Lacan, Book II*, ed. Jacques-Alain Miller, trans. Sylvana Tomaselli (Cambridge, New York, New Rochelle, Melbourne & Sydney: Cambridge University Press, 1988), 51.
43 The reference is to Jordan Peele's *Get Out* (Universal Pictures, 2017).
44 Blanchot, *The Space of Literature*, 31.
45 Blanchot, *The Space of Literature*, 31–32.
46 Foucault, 'The Thought from the Outside,' 11 & 15.
47 Michael Wood, *The Magician's Doubts: Nabokov and the Risks of Fiction* (London: Chatto & Windus, 1994), 206.
48 Harris, *Williams Burroughs and the Secret of Fascination*, 10.
49 Liana Ashenden, '*Ada*'s Erotic Entomology,' *Nabokov Studies*, Vol. 6, No. 1 (2000), 142.
50 Ashenden, '*Ada*'s Erotic Entomology,' 130.
51 Wood, *Magician's Doubts*, 230.
52 Vladimir Nabokov, *Ada or Ardor* (London, etc.: Penguin Books, 2015), 365.
53 Steven Connor, 'Fascination, Skin & Screen,' *Critical Quarterly*, Vol. 40, No. 1 (April 1998), 12.
54 Nabokov, *Ada or Ardor*, 438.
55 Nabokov, *Ada or Ardor*, 307.
56 Nabokov, *Ada or Ardor*, 292.

57 Ashenden, '*Ada*'s Erotic Entomology,' 291 & 342.
58 Ashenden, '*Ada*'s Erotic Entomology,' 129. Early on in the novel, Ada tells Van that: '*Je raffole de tout ce qui rampe* (I'm crazy about everything that crawls),' Nabokov, *Ada or Ardor*, 41.
59 Harris, *Williams Burroughs and the Secret of Fascination*, 10.
60 Harris, *Williams Burroughs and the Secret of Fascination*, 3.
61 Harris, *Williams Burroughs and the Secret of Fascination*, 10. See also Maurice Blanchot, *The Space of Literature*, 31.
62 Harris, *Williams Burroughs and the Secret of Fascination*, 9. Thanks, also, to Maxime Goergen for this unpacking of the differences between the two states.
63 Nabokov, *Ada or Ardor*, 55.
64 Edgar Allan Poe, 'The Gold Bug,' in *The Complete Tales and Poems of Edgar Allan Poe* (London, etc.: Penguin Books, 1982), 48 & 56.
65 Man Xu, Ainsley E. Seago, Tara D. Sutherland, & Sarah Weisman, 'Dual Structural Colour Mechanisms in a Scarab Beetle,' *Journal of Morphology*, Vol. 271, No. 11 (2010), 1300.
66 Nabokov, *Ada or Ardor*, 107–108.
67 Amy Ockenden, 'Seeing the Unseen: Investigating the Polarisation Properties of the Scarab Beetle Exocuticle and the Adaptive Role of reflecting Polarised Light,' Unpublished Practical Project, University of Auckland, 2 January 2017. Accessible via: https://www.researchgate.net/publication/315410121_Investigating_the_polarisation_properties_of_the_scarab_beetle_exocuticle_and_the_adaptive_role_of_reflecting_polarised_light [Accessed 19 Nov 2021].
68 Nabokov, *Ada or Ardor*, 246.
69 Nabokov, *Ada or Ardor*, 245 & 246.
70 Jacques Derrida, *Writing and Difference*, trans. Alan Bass (New York & London: Routledge, 2001), 266.
71 Wood, *Magician's Doubts*, 211.
72 Nabokov, *Ada or Ardor*, 46.
73 Nabokov, *Ada or Ardor*, 64–65, 279, 123 & 152.
74 Nabokov, *Ada or Ardor*, 42.
75 Theodor Adorno, 'Punctuation Marks,' trans. Shierry Weber Nicholson. 9 January 2019. Accessible via: http://ubu.com/papers/Adorno-Theodor-W-Punctuation-Marks.pdf [Accessed 4 May 2022], 300; Steven Connor, Lecture given at *Beckett and Animality* Conference, University of Reading, 26 September 2009. Accessible via: http://stevenconnor.com/flymean.html [Accessed 10 October 2019], 6.
76 Nabokov, *Ada or Ardor*, 205.
77 On this note, see Derrida, *Writing and Difference*, 250.
78 Paul H. Fry, 'Moving Van: The Neverland Veens of Nabokov's *Ada*,' *Contemporary Literature*, Vol. 26, No. 2 (Summer 1985), 123.
79 Nabokov, *Ada or Ardor*, 262.
80 Jacques Derrida, 'Fourmis,' in *Hélène Cixous Rootprints: Memories of Life & Writing*, ed. Hélène Cixous & Mireille Calle-Gruber, trans. Eric Prenowitz (London & New York: Routledge, 1997), 119–126.
81 Derrida, 'Fourmis,' 120; Roland Barthes, *The Pleasure of the Text*, trans. Richard Miller (New York: Hill & Wang, 1975), 4.
82 Barthes, *The Pleasure of the Text*, 20–21.
83 See Chapter 2 on Ernst Jünger and also Andreas Huyssen, 'Fortifying the Heart—Totally: Ernst Jünger's Armoured Texts,' *New German Critique*, No. 59 Special Issue on Ernst Jünger (Spring–Summer 1993), 3–23.

84 Clarice Lispector, *The Passion according to G.H.*, trans. Idra Novey (London, etc.: Penguin Books, 2014), 12.
85 Connor, 'Making Flies Mean Something,' 8 & 11.
86 Connor, 'Making Flies Mean Something,' 6.
87 Jonathan Lethem, *Motherless Brooklyn* (New York: Vintage, 2000), 7.
88 Ulrika Maude, '"A Stirring Beyond Coming and Going": Beckett and Tourette's,' *Journal of Beckett Studies*, Vol. 17, No. 1–2 (September 2008), 155.
89 Lethem, *Motherless Brooklyn*, 266.
90 Ronald Schleifer, 'The Poetics of Tourette's Syndrome: Language, Neurobiology and Poetry,' *New Literary History*, Vol. 32, No. 3 (Summer 2001), 568.
91 Schleifer, 'The Poetics of Tourette's Syndrome,' 569 & 564.
92 Schleifer, 'The Poetics of Tourette's Syndrome,' 568.
93 Lethem, *Motherless Brooklyn*, 7.
94 Sigmund Freud, 'The Psychopathology of Everyday Life,' in *The Standard Edition of the Psychological Works of Sigmund Freud*, Vol.VI (1901), trans. & ed. James Strachey & others (London: Vintage, 2001), 1–2 & 4–5.
95 Freud, *Psychopathology of Everyday Life*, 8.
96 Freud, *Psychopathology of Everyday Life*, 110 & 112–113.
97 Freud, *Psychopathology of Everyday Life*, 113.
98 A.S. Byatt, *Angels and Insects* (London: Vintage, 1995), 132, 51 & 153. [Emphasis original].
99 Colin Davis, *Critical Excess: Overreading in Derrida, Deleuze, Levinas, Žižek and Cavell* (Stanford: Stanford University Press, 2020), 181 & 168.
100 Davis, *Critical Excess*, 176.
101 Byatt, *Angels and Insects*, 141.
102 See Michelle Weinroth, '"Morpho Eugenia" and the Fictions of Victorian Englishness: A.S Byatt's Postcolonial Critique,' *English Studies in Canada*, Vol. 31, No. 2–3 (June–September 2005), 187–222.
103 Byatt, *Angels and Insects*, 3.
104 Byatt, *Angels and Insects*, 7, 49 & 60.
105 Byatt, *Angels and Insects*, 74–75.
106 Lispector, *The Passion according to G.H.*, 35 & 32–33.
107 Donna Tartt, *The Goldfinch* (New York, Boston, London: Little, Brown and Co., 2013), 44; Vladimir Nabokov, *Speak, Memory: An Autobiography Revisited* (London: Weidenfeld & Nicholson, 1967), 12–13.
108 Tommy Orange, *There There* (London: Vintage, 2018), 20.
109 Joyce Carol Oates, *The Sacrifice* (London: Fourth Estate, 2015), 245–246.
110 Charlotte Sleigh, *Six Legs the Better: A Cultural History of Myrmecology* (Baltimore: Johns Hopkins University Press, 2007), 153.
111 Gilles Deleuze, 'He Stuttered,' in *Essays Critical and Clinical*, trans. Daniel W. Smith & Michael A. Greco (London & New York: Verso, 1998), 110 & 111.
112 Blanchot, *The Space of Literature*, 51.
113 Maurice Blanchot, *The Infinite Conversation*, 46.
114 Jacques Lacan, *The Seminar of Jacques Lacan: Anxiety, Book X*, ed. Jacques-Alain Miller, trans. A. R. Price (Cambridge & Malden: Polity Press, 2015), 6.
115 Lispector, *The Passion According to G.H.*, 83.
116 Jacques Lacan, *Four Fundamental Concepts of Psychoanalysis*, trans. A. Sheridan (New York & London: W.W. Norton & Co., 1998), 75.
117 Lispector, *The Passion According to G.H.*, 24 & 72.

118 Lispector, *The Passion According to G.H.*, 65.
119 Lispector, *The Passion According to G.H.*, 182.
120 Claire Williams, '*The Passion According to G.H.* by Clarice Lispector,' in *The Cambridge Companion to the Latin American Novel*, ed. Efraín Kristal (Cambridge, UK: Cambridge University Press, 2005), 246.
121 Lispector, 'To Possible Readers,' *The Passion According to G.H.*, xi.
122 Giorgio Agamben, *Potentialities*, ed. & trans. Daniel Heller-Roazen (Stanford: Stanford University Press, 1999), 261.
123 Lispector, *The Passion According to G.H.*, 4, 52 & 184.
124 See Chapter 5.
125 Lispector, *The Passion According to G.H.*, 25.
126 Lispector, *The Passion According to G.H.*, 29.
127 Lispector, *The Passion According to G.H.*, 30 & 32.
128 On 'epidermalization,' see Frantz Fanon, *Black Skin, White Masks*, trans. Charles Lam Markmann (London: Polity Press, 2008), 84.
129 Lispector, *The Passion According to G.H.*, 49.
130 Lispector, *The Passion According to G.H.*, 36.
131 Jacques Lacan, *Four Fundamental Concepts of Psychoanalysis*, trans. A. Sheridan (New York & London: W.W. Norton & Co., 1998), 73–74.
132 Fanon, *Black Skin, White Masks*, 124, fn 25 & 117. See also Michelle Stephens, 'Skin, Stain and Lamella: Fanon, Lacan and Inter-Racialising the Gaze,' *Psychoanalysis, Culture & Society*, Vol. 23, No. 3 (2018), 310–329.
133 Lispector, *The Passion According to G.H.*, 3. See also Lacan, *Four Fundamental Concepts of Psychoanalysis*, 74.
134 Lispector, *The Passion According to G.H.*, 74.
135 Lispector, *The Passion According to G.H.*, 8.
136 Williams, '*The Passion According to G.H.* by Clarice Lispector,' 251.
137 Hélène Cixous, *Reading with Clarice Lispector*, trans. Verena Andermatt Conley (Minneapolis: University of Minnesota Press, 1990), 37.
138 Adorno, 'Punctuation Marks,' 302.
139 Lispector, *The Passion According to G.H.*, 22, 23 & 19.
140 Lispector, *The Passion According to G.H.*, 42 & 118.
141 Roland Barthes, *The Neutral*, trans. Rosalind E. Krauss & Dennis Hollier (New York: Columbia University Press, 2005), 14; Lispector, *The Passion According to G.H.*, 16 & 7.
142 Cixous, *Reading with Clarice Lispector*, 26–27.
143 Lispector, *The Passion According to G.H.*, 55.
144 Davis, *Critical Excess*, ix.
145 Neil Badmington, 'Theorizing Posthumanism,' *Cultural Critique*, No. 53 Posthumanism, Winter 2003, 11.

7 Still Life as Extinction Event

In a letter to the narrator in Bruce Chatwin's novel *Utz* (1988), Professor Orlík of the National Museum in Prague informs his recipient of his current research project: 'a study of the housefly (*Musca domestica*), as painted in Dutch and Flemish still lifes of the 17th century.'[1] He asks the narrator, an unnamed American scholar, to perform the basic information-gathering task necessary for the project, namely to 'examine every photograph of paintings by [Ambrosius] Bosschaert, [Jan] Van Huysum or [Jan] Van Kessel, and check whether or not there was a fly in them.'[2] The narrator refuses to respond, not least because Orlík is an eccentric figure, alternatively studying the extinct and colossal (the woolly mammoth) and the excessively common and mundane (the housefly), an oscillation between orders of magnitude that began with the publication, decades in the making, of his 'magisterial' article 'The Mammoth and His Parasites.'[3] Had the American scholar complied with the Professor's demand, he would have found an enormous number of flies in paintings, executed by not only Bosschaert (1573–1621) and Van Huysum (1682–1749), both of whom specialised in flower paintings, or Van Kessel (1626–1679), who frequently drew studies of insects, but also, to name but a few and in no particular order, Jan Brueghel the Elder (1568–1625), Willem van Aelst (1626–1683), Dirck de Bray (1635–1694), Jan Davidsz de Heem (1606–1684), Clara Peeters (1594?–1657/1659?), and Roelant Saverij (1578–1639).

In lieu of the unnamed scholar, I have taken up Orlík's assignation, looking for flies (and the insectile more broadly) in still lifes of the 17th and 18th centuries in art galleries I purposefully travelled to or happened to find myself in: the Wallraf-Richardz Museum in Köln, the Staatsgalerie in Stuttgart, the Städel Museum in Frankfurt (on the Main), the Gemäldegalerie in Berlin, and the Rijksmuseum in Amsterdam. As will become clear, the thinking and writing done on this matter, the result, also, of a process of incidentally discovering material, by necessity, or I should say compulsion, involve chance encounters and acts of surrender: still lifes compel such surrendering. There are a number of issues prompting my interest in this genre besides the occurrence of the insectile—addressed

DOI: 10.4324/9781003322757-7

in the second part of this chapter—beginning with the nature of the gaze, always arriving from outside and, as Lacan describes it in *Four Fundamental Concepts of Psychoanalysis* (1973), displacing the viewing subject, who keeps returning to the scene of her dispossession. One of the many curious paradoxes of the still life is that while it depicts possessions, it actually repels notions of (self-)ownership and of the 'proper' because it is an object of fascination, that is, an image that 'touch[es] you with a gripping contact,' as Maurice Blanchot notes in *The Space of Literature* (1955).[4] That which fascinates initiates and maintains contact at a distance, functions like a type of remote control if you wish or, to use Freudian vocabulary, is understood in terms of the terror of castration, as a threat to the phallic (by default white, male, able-bodied, etc.) subject. I deploy this vocabulary not to defend or keep safe this particular subjecthood but to remain aware of fascination's 'passion for the image,'[5] the gaze that keeps the I in thrall, engulfed in a field that, like its confronting gaze, constitutes an outside. It is, further, precisely the deconstruction of phallic subjectivity that I locate in still life painting, interpreted as extinction event for said subject no longer doing the looking but held captive in the pure exteriority of the other's gaze.

This is not, per se, a new approach to take, even if the still life as a historical genre is rooted to insides: inside spaces or interior settings; the ordering of space beyond these settings; the integration of things into categories determined by observation, description, classification, and exhibition—in short, the fundamental codes of a culture of objective forms. In *Looking at the Overlooked* (1990), Norman Bryson comments on the visual fields presented in still lifes, narrow but at the same time expansive, moving from real to unreal or simulated, the paintings further opening out into a 'world minus human consciousness'.[6] Hanneke Grootenboer, for her part, investigates that which is 'pending' in still lifes, governed, so she argues, by a 'more abstract or theoretical paradigm' that offers room for contemplation or, more specifically, for 'pensiveness,' a 'state that is neither active nor passive but remains in between the activity of thinking and the passivity of being lost in thought'.[7] The interiority of the image, in this case van Huysum's *Flower Still Life* (1724), has the potential to give way to an unexplored openness, initially emerging from a snapped tulip, then from a dew drop which, in its collapse of opacity and transparency, tends toward the unthought, the 'virtuosity' of a thinking made possible through this opposition.[8] I am indebted to this work, already leaning toward the outside, a concept theorised by Foucault about Blanchot's writings: the outside refers to an experience which profoundly undoes, a space which is never figured, does not cohere.

I want to locate the genre of the 17th- and 18th-century still life within the discourse of extinction narratives or, more precisely, as extinction narrative. The painter Isabella Kirkland offers one such argument with her *Taxa* series (1999–2004), still lifes executed in the realist (and magic realist) tradition that can be accessed through her website, allowing the cursor to hover and enlarge, thereby study in detail, every aspect of the paintings. *Taxa* builds a 'visual record of what we are going to lose,' writes Kirkland; her images 'stand witness' to species loss (the word 'witness' recurs in her statement: it is her duty to observe, record and mourn this loss).[9] A key identifies each species depicted, yielding a kind of paint-by-numbers kit ascertaining the particular kind of animal or plant upon which the cursor alights, its movement mimicking the process of pollination though the process is dead and artificial and seems to emphasise the deadness of the scene. Ursula Heise analyses Kirkland's paintings as databases, part of an 'encyclopaedic impulse' to determine global loss that does not, despite its act of listing, banish narrative: the logic of listing, of 'taking full stock,' participates in the elegiac or tragic mode of thinking or imagining extinction.[10] Heise observes that Kirkland combines realism and anti-realism in her compositions, showing certain birds, in 'Gone' (2004), for example, as alive, while others are figured as eggs or feathers, remnants of a presence that can no longer be summoned up.[11]

Historically, however, still lifes already combine elements that would never have co-existed, neither temporally nor spatially, meaning that the scenes they imagine have always had a certain unreality about them, a sense that they could not be witnessed as such. Indeed, I want to put the act of 'standing witness' under pressure, even erasure, presupposing, as it does, a witnessing subject or phallic I. In *Stillleben* (1997), Harun Farocki explores the historical and contemporary still life, the latter in the form of advertising images—a beer brewed in Dortmund; a Cartier watch—disappearing the labour involved in staging the objects of desire.[12] The finished product, the advert, features these objects floating in space, spontaneously, as if 'dancing of [their] own free will'.[13] More than anything, Farocki ponders the mystical character of the commodity by putting it back into relations with people at work, not in terms of an object's manufacture but as elements that need to be carefully stage-managed. Yet, at the same time, his film is drawn towards the absence of the so-called human as possibility to think the unthought, uneasily existing alongside the still life as commodity-form. His documentary, in fact, traces the concept of the unthinkable [*das Unvorstellbare*] in its transpositions into, and back out of, representation: from the notion that painting should not depict the divine (itself fundamentally unthinkable) to rendering 'human' objects as divine and therefore unthinkable, as sensuous objects unrelated to living, earthly labour. The film is narrated by a shadow, an absent narrator, and ends with a voice-over reference to the labourer who does not appear with her product, neither in the traditional still life nor in its contemporary iteration, nonetheless secretly attesting [*bezeugen*] to her existence. The observer, in turn, becomes unthinkable, her missing image—the screen by now black; only the disembodied voice remains—the starting point to a new configuration of the so-called human [*Menschenbild*]. The result of capitalist modes of abstraction disposing of this figure, its absence prompts the narrator to, curiously, considering the context of the film, posit that its disappearance is also a potentiality, the black screen, no less, the equivalent of the 'writing tablet on which nothing is written'.[14] A new *Menschenbild* might arise (or not); the still life, despite its functioning in historical and contemporary economies of desire, asserts the possibility of a thought experiment, a pensiveness in which the I no longer stands to attention but a state in which the eye, fascinated, becomes the vision of no one.

I propose, then, the following: still life is extinction event for the phallic subject by offering up 'inhuman' perspectives (1) that arise because of the structure of the gaze (covered in the first section of the chapter) and (2) which enter into the frame of reference through the insectile, specifically the figure of the fly in its function as *trompe l'oeil*. The aspect of fascination—often invoked but not analysed—is crucial to this piece and to the energy of the still life. As Oliver Harris has shown, fascination involves returns, a magnetism at once irresistibly attractive and repulsive at the same time, a materiality whose content is always elusive.[15] He further notes that the subject does not seek meaning or some rationality through these returns but finds an excruciating pleasure in submitting itself to failing to know, to repeatedly experiencing moments of surrender.[16] Lacanian thought organises my reading of still lifes, not least because of the importance of his work on the gaze and the central role that fascination plays in his theory of subject formation but also because his analysis of Hans Holbein's *The Ambassadors* (1533) serves as reference point for still life paintings in which the 'form' of the fly occurs. Anamorphosis, the 'foreign' object at the bottom of Holbein's painting, is a detail that 'sticks out' of an 'idyllic scene,' as Slavoj Žižek explains, thereby opening up 'the ground of the established, familiar signification'.[17] This detail or 'surplus knowledge' has, so Žižek continues, an 'abyssal effect on the perspective' of the viewer,[18] presided over by a stain which, despite its deforming force, springs into form as the viewer moves away. The fly, though ostensibly form, especially if rendered in early modern painting where it is enormously detailed—more generally, the delicate forms of insects fascinated artists as well as collectors and naturalists,[19] according to Janice Neri—is never neither quite form nor figure. Steven Connor notes that the fly 'does not form a figure, nor yet exactly figure a form' but that it figures 'unfigurability,' prompting reflections on form and the decaying or decomposing of form.[20]

Concepts of form and *Unform* arrange my response to the still life, as do possession and loss, fascination and displacement. Fascination is desire and, while constitutive of the form of the ego, also is a process of capture whose mechanism is one of mimicry, of losing or dispersing the coordinates between self and other. According to Blanchot, fascination is a moment of contact with the featureless, with 'the opaque, empty opening onto that which is when there is no more world, when there is no world yet'.[21] The world, in this state of suspension, falls away, and forms dissolve. Fascination is, as such, an experience of the outside that de-structures subjecthood even as it apparently assembles it. Still life, consequently, at any rate associated with the parergonal—Derrida's term for that which arrives from elsewhere—is a 'form' of address that, cutting in, beckons outside. It does so through multiple propositions operating in conjunction to each other, not least of which is that displacement of the (fascinated) I. Usually in command, always already granted representation as well as self-determination, the phallic I becomes object, taking its place among other objects, upending the usual relations in which the subject operates, structuring the world according to 'his' gaze. If still life escapes its insides and manifests instead an absolute opening in which the I loses its footing, then a different ordering of subjectivity—one which only 'retain[s] a minimum of strata, a minimum of forms'[22]—is made possible. In *A Thousand Plateaus* (1980), Deleuze and Guattari imagine an 'enterprise of desubjectification' working away at sovereign personhood to yield a constantly transforming 'minimal subject.'[23] Though in this instance talking about music, Ravel and Debussy, for example, '[preserving] a minimum of form in order to take it to its bursting point,'[24] a similar process is at work in still lifes, whose fascination and pensiveness intensify with the 'figuration' of the fly, sticking out, disturbing, preying on thought that forms and unforms itself in response to the insect's presence. The genre, then, draws the subject into arrangements—'open, without intimacy, without protection or retention'[25]—into which it disappears. This movement or 'involution,'[26] to use Deleuzo-Guattarian terminology, has an obvious ethico-political dimension, evoking, as it does, the 'form' of a minimal subject. Rather than standing witness, the I abandons itself to the curious, arresting rhythm of the still life and thereby to urgently needed new perspectives of the 'posthuman' in the current age of extinction.

In a first instance, this chapter proceeds by investigating how the mechanism of fascination works, how its various relations—strategies of arrest or suspension like glazed, gliding surfaces; moments of irruption or sudden hailing; fetish objects and indeterminate economic spaces—produce planes of desubjectification. The methodological approach is largely Lacanian psychoanalysis, structuring my readings of still life and still life scholarship, occasionally moving off into strata aligned with giving yourself over to certain works or kinds of art, including figurative painting. In this vein, the reading below is also rhizomatic, in that lines of articulation produce an essay-assemblage that can be allusive and elliptical because itself given over to ideas of dismantling and being dismantled, exteriorities that rattle the subject to and at its 'core.' The second section considers the fly as *trompe l'oeil* in its function as fascinating hinge point and parergonal element, the latter referring to both the development of the still life genre and its deployment of insects. The aim of the chapter is to take aim: to approach the still life as an event which profoundly disturbs and undoes the form or *Gestalt* of the phallic subject.

'Forms' of the Gaze

In the Staatsgalerie in Stuttgart hangs Clara Peeters's *Blumenstrauß* (c. 1612; Figure 7.1), a painting of a bouquet of flowers arranged in a vase. The vase is placed on a table, or perhaps (admittedly rather unlikely, considering the narrow strip of light at the furthest edge) it is a niche, whose curved chamber walls are just out of sight. The vase's support remains indistinct, bathed in shadows, the background empty. It contains an assortment of flowers in a highly symmetrical composition hinging on the vector of a tulip leaf arching into the air. On either side of that vector are tulip heads, their colours, white and pink, lit up against the dark background. The tulips' other leaves, at the outer edges, in turn fold towards the lower half of the painting, initiating the journey towards the bouquet's heart: eyes are guided by the graceful flexure of the bent leaves. Even while describing the scene, I find my gaze approximating the hovering behaviour of gnats, in warm weather oscillating around particular points in space, my gaze remembering the movement of apprehending a

Figure 7.1 A vase with flowers of various kinds and colours. A fly sits on the lower left-hand side of the vase. The background to this scene is empty. Clara Peeters, *Blumenstrauß* (c. 1612).

Photo: © Staatsgalerie Stuttgart

flower piece from the countless still life paintings in which insects act as surrogate viewers in a genre, after all, devoid of the 'human' figure. Here, in Peeters's work, a single fly sits on the vase on the lower left-hand side of the painting; it faces upwards, directing my eye towards the cluster of flowers (four roses, two of which are relatively large; an anemone and a carnation, all of which form part of the symmetrical axis) at the tightly packed centre. Dew drops on leaves pointing downward do not flow off but hang suspended: time has been arrested, the scene is one of utmost unreality, despite the enormous attention paid to the smallest details of the bouquet.

Arriving at the surface of this particular or indeed any other generic flower piece—largely the focus of the argument below—my gaze, even if carried and affected by surrogates, is greeted by nobody in return. The object, the bouquet of flowers, does not acknowledge my presence *qua* subject, even though it is looking at me all the same. The point of reference is Lacan's pivotal encounter with a sardine can floating on the water, in the sun, off the coast of Brittany. 'Witness to the canning industry,'[27] which Lacan and the fishermen he joined in labour were meant to supply, the sardine can does not heed the subject in its function as subject. The scene Lacan relays in *Four Fundamental Concepts* is one of non-belonging, already evident in his position as a young intellectual amidst the preterite. Lacan admits to being 'out of place in the picture,'[28] but the context is not only one of class privilege on the one hand and preterition on the other, even though that might be the most striking aspect of the encounter. It also concerns the moment where the subject falls, comes under the unfathomable gaze of the other. The sardine can episode reflects on the elision of the subject as such in a visual field—Lacan calls it a play of light and opacity—and of being grasped by the thing that looks at it. The sense of non-belonging describes the most essential psychic reality of the subject, whose representations belong not to the I but to the other, holding the I in its gaze.

In *The Sight of Death* (2006), T.J. Clark recalls the circumstances that led him to write about Poussin's *Landscape with a Man Killed by a Snake* (1648) and *Landscape with a Calm* (1650–1651), which were facing each other at the Getty Research Institute in Los Angeles, where Clark spent the early months of the year 2000 as part of a fellowship granted by the National Endowment of the Arts. He had seen *Landscape with a Man Killed by a Snake* repeatedly while living in London in the 1960s without considering it in its significance as a painting that, for him, '[sums] up the utmost that visual imagery could do in a certain vein'.[29] His book is the result, precisely, of an incident or accident deeply bound up with repetition, because the subject matter, Poussin's two paintings, can neither be 'retained in the memory' nor be 'fully integrated into a disposable narrative of interpretation'.[30] Clark's project resonates, not only because of the compulsive, as well as chance, elements of its emergence and development, the sheer force the paintings exerted on him, his unexpected response and the form of his response to coming across them at the Research Institute while looking for nothing in particular. His project attests to the power of being held in the gaze of the other, which always and totally exceeds the capacity of the eye, memory, the processing systems of the viewing subject.

In many ways, *The Sight of Death* reads like a Lacanian case study, considering how Lacan presents the effects of the gaze, whose characteristic features he borrows and develops from Sartre. The gaze is something which the subject is confronted with by surprise, as if by turning a corner and, without warning, being hailed. Not necessarily linked to the organ of sight, the gaze occurs, for example, like the 'sound of rustling leaves, suddenly heard while out hunting' or like 'a footstep in a corridor'.[31] It is this configuration—Lacan, the sardine can, Clark's (in)attention, and the sudden confrontation with the other's gaze—that shapes my reaction to the still life, whose object arrangements, with or without insects, have on each and every occasion provoked a profound disturbance. Lacan famously talks about Holbein's *The Ambassadors*, in which two figures stand frozen, 'stiffened in their adornments'.[32] The scene, otherwise so choreographed and expensively arranged, is unsettled by a 'form … What? A skull',[33] whose very notation, the ellipses, the interruption of the question 'What?' speak to its deforming substance. It is as if another dimension was opening up at the figures' feet, where the 'form' intrudes:

> What, then, before this domain of appearance in all its fascinating forms, is this object, which from some angles appears to be flying through the air, at others to be tilted? You cannot know—for you turn away, thus escaping the fascination of the picture. Begin by walking out of the room in which no doubt it has long held your attention. It is then that, turning round as you leave, [… that] you apprehend in this form … What? A skull.[34]

Lacan argues that what Holbein makes visible is the mechanism of vision which, like the constitution of the subject, must repress 'forms' of the 'real'. The anamorphic ghost is an object in 'geometral' space, an object reproduced in an oblique position, located in a visual field different from the one the subject is occupying at that moment. The movements effected by the viewer Lacan mentions—she turns away to walk out of the room and only then, in the process of turning, apprehends the form as skull—suggest as much. It might well be that the original frame of *The Ambassadors* included a space for a 'telescopic viewing device,' through which the ghost-object would have snapped into perspective.[35] Yet, Lacan adds, the ghost is 'the gaze as such, in its pulsatile, dazzling and spread out function,'[36] disorganising the fascinated subject's field of vision. It is the 'form' of the gaze as *objet a*, the function of lack, inscribing the subject in the grid of desire. The domain of vision, hence, becomes integrated into the field of desire, which is the field of the other, of the 'seeing' object.

Still lifes enact death; *nature morte* in French, the term itself is caught in a relation of suspension, a structural ambivalence that goes way beyond the opposition between life and death. For Harry Berger Jr, floral still lifes, especially, are a 'genre devoted to moribund organisms,' considering that the objects they depict are already dead and in the process of decaying; the bouquets, despite the vibrancy of their colours, their full blooms, a '*nature mourant[e]*'.[37] De Heem's *Festoon of Fruit and Flowers* (1660–1670) at the Rijksmuseum is especially striking with regard to this fundamental ambivalence. It seems to sum up my disquiet vis-à-vis still lifes, whose beckoning power has always been bound up with being hailed, arrested, trying to escape their gaze, turning my back and turning back to, once more, be turned to stone by their force of captivation. The 'form' at the central point of de Heem's painting is a pomegranate whose husk has burst and whose interior wall can barely hold the seeds contained in the fruit's cells. Right below, an ornamental pumpkin, whose outer shell is a landscape of the smooth and the 'warted,' the bulbous protuberances, like an infestation or disease, interrupting the surface of the skin.

Upon closer inspection, the painting, more generally giving the suggestion of bursting, of being obscenely thrust into the field of vision, reveals further agitations that irrupt from ground zero of the pomegranate, initiating a series of epidermal events. *Festoon* is an experiment in the variation of representing texture—the texture of surfaces and skins—whose affective experience engenders an interplay of desire and disgust, a sensation of intolerable enjoyment. On the downy skin of a peach next to the pomegranate, a housefly, the device by which the gallery visitor's eye moves from one insect to another, gradually wandering along the perimeter of the arrangement. We can discern what looks like an earwig, nearly swallowed up in the dark background on the lower right of the ornament. At the lower left, on a leaf whose veins glow against the darkness, sits a darkling beetle of some kind, its back catching the light. The overwhelming impression, here, is one of excess, while the form of attention in operation is that of fascination, capturing the viewing subject in a state at once alluring and potentially destructive. It recalls, of course, Freud's reflection on Medusa's head, turning the spectator stiff with terror, the gaze of the other rooting the phallic subject to the spot. Fascination is indeed the word that recurs in relation to still life paintings; Hanneke Grootenboer mentions it in *The Rhetoric of Perspective* (2005), where she notes that the genre 'is a product of Dutch seventeenth-century culture and its expressed fascination with observing and recoding reality in a variety of circumstances'.[38] Elizabeth Alice Honig, though talking about Dutch marketplace paintings, argues along similar lines:

> Seventeenth-century Dutch genre painting, at its best, derives its unsettling power from its transfiguration of the commonplace. Subjects often banal in the extreme are removed from the ordinary context of experience and, perpetually fixed by the painter's deft brush, transfix our gaze to become the objects of an odd fascination, a fascination seemingly inappropriate to their original nature. Recognition—'this is a world like our own'—disturbingly plays off the image's absolute removal, as a work of art, from that world [...].[39]

Hans-Joachim Raupp, in *Stillleben und Tierstücke* (2004), describes the still life's 'floating' between artificiality and verisimilitude as the source of its *Faszinationskraft*. Svetlana Alpers, without using the term, nonetheless describes its condition, the rapt attention to a world of surfaces, the picture itself 'as surface (like a mirror or a map)'.[40] The fascination of the gaze—there, also, in Freud's analysis of Leonardo Da Vinci's Gioconda—involves that 'absolute removal,' a *Realitätsverlust*, from the world around the transfixed subject, who surrenders to the object's enigma.

Given the minutiae of its depicted objects, the result of what Alpers calls an 'attentive looking, transcribed by hand',[41] the Dutch still life—meaning, above all, the flower and *pronk* pieces, both of which are intent on displaying wealth—renders commodities, fantasics of affluence, and is itself a commodity. Miya Tokumitso, for example, uses Willem Kalf's *Still Life with Ewer, Vessels and Pomegranate* (mid-1640s) to demonstrate the costs necessary to produce the painting, whose ultramarine pigment alone was exceptionally expensive, deriving from the semi-precious lapis lazuli stone imported to Europe from Afghanistan via Venice.[42] Still lifes and their relation to accumulation or as acquisition is an argument that is often made, for obvious reasons: there is the matter of their historical emergence, linked to a culture of collecting, the *Wunderkammer* or curiosity cabinets that gathered and put on show objects removed from their contexts to exist solely as objects. Roland Barthes notices that the painted things are always open: the sliced and half-peeled lemon, 'caught at the precious moment it exchanges the scandal of its perfect and useless ellipse,' curling down the table, 'for the first of its economic qualities, astringency'.[43] Barthes's observation can be supplemented or confirmed by specifying other instances, the pomegranate and fig in *Festoon of Fruit and Flowers*; a pomegranate, a walnut and a melon, weirdly soft and shapeless, as if melting, in Van Huysum's *Still Life with Fruit and Flowers* (c. 1728); the bread and oysters, not to mention the lemon in its position at the edge of the table, in Willem Claesz Heda's *Still Life with a Gilt Cup* (1635). These objects, Barthes continues, destroy themselves as closed substances,[44] thereby forming part of a great taxonomy of exposure. The logic of visibility doesn't stretch to encompass the relentless operations of the market, however, whose magic, masking a universe of pain and exploitation, brings them to the Dutch table.

The argument about still life as a non-narrative genre thus makes sense only up to a certain point. As Julie Hochstrasser has shown, the 'obdurate silence' of the (often costly) objects can be counteracted by insisting on their provenance, their histories of movement, the social costs of their deployment opposing their 'easy surfaces'.[45] As Barthes has shown elsewhere, objects that are 'glossed over'—naturalising them: they are to be passed over in silence and exist innocently, eternally—are frozen into these 'easy surfaces,' where history disappears into pure form.[46] In the context of the still life, Barthes is thinking of the 'defining power' of the Dutch canals, calling the whole process of assemblage a 'water-merchandise complex,' by which he means as much a particular aesthetics of the glide than a technology of accumulation: 'it is water that makes the object, giving it all the nuances of a calm, planar mobility.'[47] He remarks on the glaze that coats objects, lubricating the gaze: opened up but no less of a surface, they 'glance' in the direction of the viewing subject. A mythological system is at work, abolishing complexity, yielding a world without depth, and concurrently setting the scene of fascination and fetishization. The glance, after all, *viz.* Freud, is endowed with light, objects shining forth according to the logic of possession but also of traumatic loss.

Hal Foster, focussing on *pronkstilleven*, writes that the still life's strange energy—Honig, like Raupp transfixed by the genre's oscillation between states, calls it unsettling and disturbing—is linked to fetishistic projection.[48] The sheen (*Glanz*) of the depicted objects is significant, initiating an inversion of the subject–object relation. The banal things, removed from their ordinary context of experience, draw the light while the viewing subject, to refer back to Lacan's terminology, is plunged into obscurity. Objects, Foster continues, seem 'endowed with life to the degree that the viewer is sapped of it,'[49] a clue, perhaps, as to why the American scholar refuses to carry out Orlík's assignment. To look for houseflies in still life paintings means to suffer the event of castration, bearing in mind the genre's strategies of absorption—the gliding surfaces, the contextless objects, the total absence of world and of the 'human' organism—causing the subject-spectator to lose face and/or head. The scholar risks the loss of command, the position of the subject doing the looking, because his image is not returned: pinned down by a gaze, subject becomes 'unmanned' object. According to Foster, the Dutch still life, on the one hand, 'is perfect—perfectly composed and perfectly finished—with no hint of lack or loss'; on the other, its 'very perfection [...] points to a pre-existent loss or castration in the subject—which is precisely why perfection is demanded'.[50] Freud observes that the fetish, like the still life, is an ambivalent structure, a substitution for a loss/castration but also the symptom of immense suffering, a reminder of, and memorial to, that loss. The fetish is designed to prevent loss, or the recognition and memory of loss, but it is at the same time its 'half-way' affirmation; the *Glanz* it radiates out is intended to be triumphant, a protective structure, all the while being indicative of dispossession.[51] In this vein, the 'luminous shine' of still lifes 'recalls our lack even as it distracts us from it,'[52] Foster concludes, gesturing towards Lacan's gleaming sardine can looking back at us. The object's look, the returned gaze, reveals the commanding subject's displacement and deprivation.

Foster excludes flower pieces in his analysis; he argues that the *pronk* still life, most evidently, engages with the expanding market by fetishising objects deriving from other cultures (porcelain from China; glass from Venice; rugs from the Middle East; Nautilus cups made from shells found in the Western Pacific, etc.). Flower paintings, too, accumulate wealth from abroad and negotiate this wealth with a similar approach or finish. In *Looking at the Overlooked*, Bryson points out that flower pieces are not pastoral but georgic, all work, the result of various different forms of (acknowledged and unacknowledged) labour converging on the canvas. The 'centripetal' vase-space, accordingly, gathers flowers that bloom in different seasons and different regions of the colonial network, and is often the result of combining several separate studies, yielding an arrangement that never existed nor could ever exist.[53] Bryson reads a refusal of both space and time in these bouquets: the *nature mourante*, produced as 'natural,' is a synthetic thing, a form of dead capital. It serves to bring to prominence the enormous power of *technē*, a whole process of operations capable of transforming all dimensions. He sees a Faustian ambition at work, the genre setting out to exalt the phallic subject despite its absence from the painted scene. Yet the trajectory of his chapter consists of moving from prosperity, harmony, order, the production and accumulation of value to waste and lack, the hollowness of value and the deadness of capital, laying waste.

If the bouquet—for example, Peeters's *Blumenstrauß* (Figure 7.1)—stands in an indeterminate space, a 'no-man's land'[54] as Grootenboer comments, it reflects the economic space responsible for its existence. The returned gaze, as such, and despite all the levels of mastery and subjection that determine a painting's coming into being, also emanates from there, that is, nowhere. Bryson, discussing Kalf's opulent and disturbing *pronkstilleven*, notably *Still Life with Nautilus Cup* (1662), finds that they tend to provoke the question as to who is to see them, is to receive these images of extravagant wealth.[55] The paintings, operating according to not only the logic of the market but also the economy of the supplement—the original object is never quite enough to guarantee its value—form part of collections, as items of trade and investment so that the table with its goods, usually a 'tactile, domestic space,' becomes abstracted: it functions like a 'bank-vault—or graveyard'.[56] These are 'inhuman' dimensions, deep and dead: wealth locked in, luxury objects that have nothing to do with the living. Although aware of the different aspects of each still life genre (the *pronk* or flower pieces; the humbler breakfast paintings),

Bryson, no doubt also thinking of the still life as a genre of things at rest, a '*nature reposée*,'[57] observes that, in each of them, objects reveal their autonomy, exist independently of any viewer. Who is to see all this is a question that cascades: What is the nature of 'my' gaze, as I stand in a museum looking at a still life? What exactly is it that looks back at me, from that space over there, commodified and dimensionless? What is the gaze that turns me into object of possession, before which I become opaque and disorganised? What, finally, is the function of the flies we encounter in the flower pieces? Are they stabilising agents, holding platforms, supplements whose 'lubricated' surfaces seek to guarantee the value of a painting? Are they, therefore, reminders of lack or markers of formlessness? Are they pensive 'figures' destroying systems of form, the phallic *Gestalt* of the subject, absorbed and in its absorption orientated toward the *tout autre* outside, and within?

Trompe l'oeil and *Passe-mureille*

Symbolic or naturalistic (as if they were mutually exclusive), insects integrate into that taxonomy of order producing and immobilising 'nature' through what Janice Neri calls a 'specimen logic,' characterised through the familiar method of isolating and decontextualizing things, whose edges, contours and appearance—in other words, their form and materiality—are 'stubbornly insisted on'.[58] Onlookers and surrogates, occasionally providing a map or stepladder of sorts by which to navigate a bouquet or festoon, insects appear as *trompe l'oeil*, duping viewers into brushing them away. A fly sits on the vase in Peeters's *Blumenstrauß* (Figure 7.1), on the vines of partially translucent grapes in van Huysum's *Still Life with Flowers and Fruit*, on a peach next to the burst pomegranate in de Heem's *Festoon*, whose opened, exploded form is indicative of the genre's entropic turn or fixation. *Trompe l'oeil* includes the plates, lemons, knives balanced precariously at the edges of the table, demonstrating, as Bryson argues, an 'interest in the ruined form,' further evident in the erupting foodstuffs 'showing material which has lost its place.'[59] *Trompe l'oeil* is allied to ruin, then, if it is also a manifestation of the artist's skill, the accurate reproduction of an object so that it seems to be fully present in three-dimensional space. The gesture of brushing away a fly, as Norman Land analyses with respect to the Renaissance painter Carlo Crivelli's work, is an act seeking to protect, to keep safe from ruin, occurring prior to recognising the device as illusion, emblematic of exquisite specimen rendition.[60]

Border phenomenon, the fly is provocation and 'hinge point,' hovering between the singular and the horror of plurality, the minimal and the maximal.[61] In *Utz*, the Professor spent 30 years researching the woolly mammoth before turning his eye to *Musca domestica*, which, at the point of meeting the narrator, he was tracing in the Prague Metropolitan area.

According to his friend Utz—who is enthralled by Meissen porcelain and could distinguish pieces deriving from clay found in Colditz or the Erzgebirge—Orlík could tell the difference between a fly that 'came from Malá Strana or Židovské Mesto or from one of the garbage dumps that now encircled the New Garden City'.[62] Chatwin's novel plays on incongruous juxtapositions, in terms of both scale (mammoth/fly) and value assignation (fly/Meissen porcelain), but there are nonetheless some points of correlation, not least in relation to the notion of the colossal. Disproportion is always established with reference to the 'man-standard,'[63] determining the good form and 'appropriate' dimensions; the colossal (like the sublime) is that which defies 'human' measure. Even though it is the wrong adjective to use in connection with the fly in its singular form, the anti-form of the swarm of flies might well prompt the colossal as an apt description: the swarm, after all, exceeds all ('man-made') measure as well as the concept of form. But the fly, even singular, is prodigious in that it surpasses all limits: it has spread all over the world and moves, back and forth, from the living to the dead and the excremental, embodying, as Steven Connor writes, spatial and categorical disturbance.[64]

Orlík confesses to 'being enchanted by the vitality of the fly,'[65] itself a rather unusual focus, considering that the fly tends to be interpreted, and behaves, as vector of disease and death, as indicator of decay. Wherever it alights, the object is instantaneously tainted—hence the gesture to brush it away—and, despite its omnipresence, it remains 'our' total other, 'entirely intractable' to 'our' will and purpose.[66] Indeed, for Orlík its 'vitality' is political, though his statement—made in a Prague restaurant, surrounded by Party Members keen to overhear critiques of the state—is clearly intended as incitement:

> It was fashionable among his fellow entomologists—especially the Party Members—to applaud the behaviour of the social insects: the ants, bees, wasps and other varieties of Hymenoptera which organised themselves into regimented communities.
>
> 'But the fly,' said Orlík, 'is an anarchist.'[67]

Other insects, Orlík thinks, can be instrumentalised, recruited to colossal state power, but the fly, vitally anarchist, fascinates because it is parasitical as well as pure libido: it interrupts, upsets, irritates, is constantly elusive. It is in and of itself a provocation, and it is as such, presumably, that he would have approached its appearance in still lifes. We don't know much about the Professor beyond his research interests; he remains a fairly marginal figure in the novel, evidently more interested in its titular character as well as in the art or psychopathology of collecting and ownership. Yet we can assume that the fly interests him more than as memento mori, taking into account that it is the fly's irritating liveliness—recall the subject of his article, 'The Mammoth and his Parasites'—not its (symbolic) death-function that impresses itself on him. He also studies the flies' various forms and variations of form; otherwise, he would not be able to classify them as occupying particular geographical locations, their iridescent wings, examined under a microscope, revealing particular class structures associated with these areas.

This starting point—parasitic figure with inscribed membranes—is certainly compelling, though it is the form of the fly, rather than its 'content' or markings, that is most pertinent for a study of *Musca domestica* in Dutch and Flemish still lifes of the 17th and 18th centuries. Etymologically, as Derrida reminds us, insect derives from *inseco*,

> which means to cut, to dissect, at times to tear with the teeth [...], to put into small pieces. The frequentative *insector* means 'to pursue without respite,' to be on the heels, to hasten energetically, to seduce, perhaps to court, to harass, to go after—etc.[68]

The fly cuts in. It is difficult to perceive it as anything other than an intruder, though it might well belong to the system, to borrow Michel Serres's words: the fly is the noise, the corruption on the heels, at the threshold.[69] Carlo Crivelli's *Madonna and Child* (1470) is case in point; Land remarks that Renaissance viewers of the painting, by brushing away the fly, were casting aside the devil, the lord of the flies, with the same gesture. Even the location of the fly (a *trompe l'oeil*) troubles: from one perspective, it seems to sit on the surface of the painting; from another, it is present in the painting, on a ledge, towards which the Virgin looks and tilts her head. Land further notices the size of the insect, too large compared with the Virgin and child—the fly is as big as the child's foot—thereby linked to the colossal, phallic fruit hanging overhead. He uses the word fascination to describe the effect the painting has on its viewers, a term deployed lightly here, as if a synonym for interesting or charming, stripped of its danger or potential. Further, curiously enough, Land places the painting's source of fascination not with the fly, in its position at once outside and inside, but in the 'provincial, perhaps eccentric presentation of a conventional subject matter'.[70] Fascination, Land implies, is a 'considered response and is extended in time,'[71] whereas the fly is the first thing we notice: it registers like a shock, an event producing disorder and confusion, both in terms of perspective and concerning the narrative, or system, of the painting.

Land is right about time stretching on (yet unconsciously so) in a state of fascination but not about it being a 'considered response,' which is characterised by deliberate thought. A fascinated subject, on the contrary, sinks into time, surrenders to its absence: time passes without notice. To be fascinated means to be detained under the irresistible influence of the other, whose gaze, as Blanchot notes, 'robs us of our power to give sense'.[72] Land, while correctly identifying the fly as the first thing we see, unthinkingly assigns the event of fascination—not really identified as such, that is, as a state of being seized—to the painterly execution of the subject matter, which plays its part. Yet it is the fly, more than anything else, that keeps us in thrall: it is the vector of the Virgin's gaze that lands there too, and the child's head is similarly turned in its direction. Were we to trace diagonal lines passing from the Virgin's eyes, her slanted head and the inclination of the child's head, their point of convergence would be the fly, looking up, as if returning the Madonna's gaze. While Land is attentive to the fly's position, forming the crux of his reflection, it warrants further investigation in its function as fascinating object, parasitic insertion or arrival, and as a kind of *passe-mureille*, in the sense that it exists here, on the outside, and there, inside the enclosure of the painting.

The fly is the detail that sticks out. In the Master of Frankfurt's *Portrait of the Artist and His Wife* (1496), a fly sits on the woman's white headdress, another by a plate of cherries, feeding off fruit that has fallen off the plate; here, too, both flies are out of proportion. The fly on the headdress, stark against its white background, resembles a stain, protruding out of the canvas: it is a *trompe l'oeil*. We feel once more compelled to brush it aside; excessively large in its dimensions, in its insistence, there, on that clean fabric, it breaches the limits of tolerance. Lacan wonders what attracts us in *trompe l'oeil*, and concludes that because the representation does not change with a shift of perspective, it is something other than it seemed, namely *objet petit a*. The question of *objet a* is a question of the gaze of the other, reducing the subject to a stain, a theory which Lacan develops by way of Roger Caillois's work on mimicry, investigating the phenomenon in certain non-human animals as linked to the dispossession of the schizophrenic. An insect, for example, disappears to merge with its environment and manifests, as such, the function of the subject as stain, vanishing into the surroundings and becoming one point among others. This process, an 'assimilation to space,' happens in 'legendary psychasthenia' too, the name Caillois uses to define a disturbance in the relations between subjectivity and space. The disturbance occurs as follows: the 'psychasthenic' subject, losing her bearings, is permeated by darkness, which is 'filled,' enveloping and passing through her; she is, in turn, transformed into 'dark space'.[73]

The Lacanian stain traces back to this dark space, a subject 'filled,' becoming mottled against a mottled background.[74] It is a term and phenomenon deriving from an event—'adaptation of form to form'[75]—defining the relationship between a mimetic insect and its *Umwelt*; the determining force, as Caillois shows, is fascination. In this drama, the organism is 'no longer the origin of the coordinates, but one point among others; it is dispossessed of its privilege and literally no longer knows where to place itself'.[76] While Caillois describes the phenomenon as a schizophrenic disorder, Lacan identifies it as fundamental to the experience (however repressed) of the neurotic subject. The phenomenal dimension of mimicry, of becoming stain, is that which generally characterises the domain of the subject, a domain that isn't hers at all. At the point of encounter—between the gaze of the other originating in space and the subject only ever integrated into, but not herself the origin of, the visual field—the subject appears or disappears in the picture as stain. According to Caillois, the mechanism functions like a 'gigantic phagocytosis' digesting the subject. In this occupation by space, the subject tries to look back at herself but finds nothing: she sees only dark space, the matter she herself has become.[77] For Lacan, this moment of encounter, absolutely determined by fascination, is provoked by *trompe l'oeil*, initiating that phagocytosis in which the gaze takes over, triumphs over the eye or the subject. In the wake of falling under the spell of the gaze, the subject is 'presented as other than [she] is, and what one shows [her] is not what [she] wishes to see'.[78]

To encounter flies as *trompe l'oeil* in Peeter's *Blumenstrauß* (Figure 7.1), de Heem's *Festoon* and Brueghel's *Still Life with Flowers in a Glass* means to confront that which usually eludes the subject thinking itself at the centre of perspective:

> If the function of the stain is recognised in its autonomy and identified with that of the gaze, we can seek its track, its thread, its trace, at every stage of the constitution of the world, in the scopic field. We will then realise that the function of the stain and of the gaze is both that which governs the gaze most secretly and that which always escapes from the grasp of that form of vision that is satisfied with itself in imagining itself as consciousness.[79]

The stain, disturbing the subject's field of vision, is that which lies beyond appearance, constitutes the other of appearance. There are several aspects to this argument, though: insects in still lifes, as we have already seen, are also enablers, in that they mobilise the eye in traversing a painting without faces. They operate as stoppage points, props to anchor, then release, then to provide another holding platform for the eye of the viewing subject, who, for once, does not see 'his' own face reflected. Looking at Heda's *Still Life with a Gilt Cup*, Grootenboer wonders about perspective in a genre in which things take precedence over the representation of the 'human;' things which, as Bryson observes, have their own distinctive, 'slow, almost geological' and entropic, rhythm.[80] Rhythm is one thing, the repetition of these fascinating forms another—they endure and decay—but it is also worth thinking about still lifes as existing in opposition to portraiture and historical paintings. Both of these genres revolve around the 'human' subject, 'his' faciality and apparently significant contributions, but which is radically without representational place in still life, further giving the impression that it has no place beyond it either. Bryson's earlier question—Who is to see all this?—is echoed in

Grootenboer's analysis, which proposes that still lifes should be understood as theories of vision, a 'form of thinking in visual terms'.[81] One of the examples she gives to illustrate her point is a reading of *Still Life with a Gilt Cup*, a rather chaotic breakfast piece arranged, as is so often the case, in 'outer' space (a featureless void), all the more remarkable because contrasted to the disorder of the table:

> Heda presents this scene we normally only register vaguely, from the corner of our eye, from a rather unusual point of view. While we see the food from nearby, it is not as potential diners and what it offers us is certainly not a feast for the eye. It is not entirely clear from what imaginary spot this picture must be viewed; indeed, it looks as if nobody will ever think it worth a glance. Yet we are neither sitting at the table nor standing in front of it, but rather 'approaching' it. While we are offered an overview of the table as if it were a landscape, the proximity of the objects and their many details disrupts this notion of having an overview. Neither a bird's- nor a frog's eye view—neither occupying a diner's position nor being offered the viewpoint of a fixed viewer—our looking is suspended at an angle relative to the tabletop that we normally would never experience. Persisting with animate metaphors, we could say that we view the table from a fly's point of view.[82]

There are a few things to comment on, here, the first of which is the lateral drawing nearer, effectuated out of the corner of the eye, barely glancing in the direction of the assemblage. This approach has much in common with still life as a genre, uninterested in the acts, dignity or prestige of a particular 'human' subject and effectively extinguishing history, especially considering the long duration of the depicted forms.[83] Still life is rhopographic: it is interested in the things apparently without importance that are usually overlooked, the material base of life and the 'prime objects'[84] taken for granted by those with means, in uneventful times. Bryson situates the matter of still life as belonging to the '*aevum*, time that has a beginning but no end,' that its forms are 'in a sense unconscious,' in that they don't need to be thought consciously, that their usage is passed on and down, without conceivable end.[85] And yet they hint at end-times or, rather, a world without the 'human' subject and entirely given over to its 'inhuman' dimensions, where the marginal (speaking from the position of the privileged subject) asserts its autonomy.

Still life is concerned with, and arrives from, the margins. Before developing into a genre in the late 16th century, symbolic arrangements of fruits or flowers were often painted on the backs of portraits or the back panels of religious scenes, on diptychs which, when closed, reveal a vase of flowers. They also occupied the fringes of paintings, were situated in the foreground when the subject of the painting—like in the Master of

Frankfurt's *Portrait of the Artist and His Wife*—took up position elsewhere; small bouquets were seated in niches, at the edges of the canvas.[86] Victor Stoichita, analysing the trope of the niche, writes that it allowed different levels of representation to come into play in a painting, and shows that the still life in a niche—a topos in religious paintings especially—evoked a multi-dimensional spatiality, depth of field or different 'insides' pressing up against each other. He likens the function of the niche to the relationship of the frame to the painting, noting that 'the frame belongs to our world, but the painting is an opening into another reality'.[87] The object of the still life in its niche, on the reverse side of paintings, in the corners or neglected foregrounds of biblical scenes, in other words at the margins of the visual field of paintings, presents a parergon to the principal work, the ergon. The parergon, Derrida writes, is neither work nor outside the work. Rather, parergon, the supplement or remainder, comes up against the ergon, from a 'certain outside,' behaving 'like an accessory that one is obliged to welcome on the border, on board.'[88] There, at the border, an 'accessory' or ornament, something 'exterior to the proper field'[89] turns up, not to be turned away: 'one is obliged to welcome' it, though it might be barely tolerable. (Tolerance, of course, always presupposes some border or limit, which must not be crossed.)

It is this dynamic—the parergon outside an ergon which, by default, requires a lack inside, some missing thing in the work to be supplemented—that structures the still life as a genre born, as Stoichita notes, 'of a cut.'[90] Its emergence marks the emancipation of the parergon becoming ergon. Still life is that which arises, to refer back to Derrida's argument, from a 'certain outside,' beyond 'the proper field' to establish itself as 'proper,' all the while bearing in mind the incomprehensibility of zones like the proper, the limit, the border. The parergon—existing outside yet within, where it is 'welcomed' or tolerated as part of an ergon's expression—is pushed forward, into 'our' vision: the still life no longer as 'hors d'oeuvre,' on the periphery of the centre of attention, but principal subject.[91] Once there, it enables and activates a vision that isn't proper, that is, which does not belong to the subject, absent, out of place, without reflection, like a vampire. Looking at a still life means to see or think a world without the 'human,' a proposition that, extrapolating from Grootenboer's reading of Heda's breakfast piece, proceeds as follows: that which is usually outside or obscene (the parergon) becomes the scene to behold, disturbing 'our' vantage point. She describes the unsettling of perspective, the mechanism of displacement: we are neither sitting at the table nor standing in front of it, and we only ever seem to advance towards it, around it, without overview or fixed position. 'Our' looking is suspended at an angle we would otherwise never experience, that is, the angle of a fly, the vehicle of a vision or thought from the outside.

What compounds all this is the marginality or 'parergonality'[92] of the fly itself: as *trompe l'oeil*, it is always already 'outside-the-work,'[93] never mind, even, their recurrence as figures at the lower margins of a painting, or half disappearing, their 'face' cut off behind the bouquet like in Dirck de Bray's *Flowers in a Glass Vase* (1671). If the prefix para, which according to Serres, 'is on the side, next to, shifted'[94] or, as J. Hillis Miller writes, behaves as 'a guest to host'[95] (or vice-versa), then the fly, offering the viewers an *Anhaltspunkt*, a possibility of ordering and stabilising their gaze, also mediates its undoing or disorder. Para 'is not on the thing, but on its relation.'[96] As such, it determines a way of thinking about systems of relations, a form of thought putting to work that which leads astray, sets free an 'internal' indetermination. In his meditation on Blanchot and thought from outside, Foucault observes that 'it is extremely difficult to find a language faithful to this thought,' perpetually in danger of being mapped back to a centre of unification, the 'dazzling interiority' of the phallic subject, repatriating such thought to its spine, verticality, the command of its gaze.[97] The still life, as I have tried to argue above, cuts in to beckon outside, bearing in mind the weird placeless places where it is situated, in which the 'human' face has vanished, the 'impossible' perspectives and ruined forms, the parergonal functions that disorganise the experience of the subject as central. If we return, further, as we are bound to do, to the series of propositions raised by the *trompe l'oeil* or *passe-mureille* fly, its 'form' indeed performs a kind of un-thinking, determined by the compulsive returns and the fascinating relation between subject and object. It is there—nowhere, in a zone of total suspension—that the privileged phallic subject, arrested and mesmerized, becomes minimal.

Notes

1 Bruce Chatwin, *Utz* (New York, London, Victoria, Ontario & Auckland: Viking, 1988), 117.
2 Chatwin, *Utz*, 117.
3 Chatwin, *Utz*, 34.
4 Maurice Blanchot, *The Space of Literature*, trans. Ann Smock (Lincoln & London: University of Nebraska Press, 1982), 31.
5 Blanchot, *The Space of Literature*, 31.
6 Norman Bryson, *Looking at the Overlooked: Four Essays on Still Life Painting* (London: Reaktion, 1990), 143.
7 Hanneke Grootenboer, 'The Pensive Image: On Thought in Jan Van Huysum's Still Life Paintings,' *Oxford Art Journal*, Vol. 34, No. 1 (2011), 15 & 17.
8 Grootenboer, 'The Pensive Image,' 29.
9 Isabella Kirkland, 'About,' https://www.isabellakirkland.com/about [Accessed 22 October 2021].
10 Ursula K. Heise, *Imagining Extinction: The Cultural Meanings of Endangered Species* (Chicago: University of Chicago Press, 2016), 65, 61 & 12.
11 Heise, *Imagining Extinction*, 80.
12 Harun Farocki, *Stillleben*, Movimentio Berlin, 1997. Accessible via: https://www.youtube.com/watch?v=KQkDUnVvryI [Accessed 27 October 2021].
13 Karl Marx, *Capital*, Vol. I, trans. Ben Fowkes (London, etc.: Penguin Books, 1990), 164.

14 Giorgio Agamben, *Potentialities*, trans. Daniel Heller-Roazen (Stanford: Stanford University Press, 1999), 245.
15 Oliver Harris, *Williams Burroughs and the Secret of Fascination* (Carbondale & Edwardsville: Southern Illinois University Press, 2003), 17.
16 Harris, *Williams Burroughs and the Secret of Fascination*, 17.
17 Slavoj Žižek, *Looking Awry: An Introduction to Jacques Lacan through Popular Culture* (Cambridge, MA & London: October Books, 1991), 90.
18 Žižek, *Looking Awry*, 90.
19 Janice Neri, *The Insect and the Image: Visualising Nature in Early Modern Europe, 1500–1700* (Minneapolis: University of Minnesota Press, 2011), ix & xi.
20 Steven Connor, *Fly* (London: Reaktion, 2006), 32.
21 Blanchot, *The Space of Literature*, 33.
22 Gilles Deleuze & Félix Guattari, *A Thousand Plateaus: Capitalism and Schizophrenia* (London & New York: Continuum, 2004), 298.
23 Deleuze & Guattari, *A Thousand Plateaus*, 298.
24 Deleuze & Guattari, *A Thousand Plateaus*, 298.
25 Michel Foucault, 'The Thought from the Outside,' in *Foucault/Blanchot*, trans. Briam Massumi (New York: Zone Books, 1987), 27.
26 Deleuze & Guattari, *A Thousand Plateaus*, 298.
27 Jacques Lacan, *The Four Fundamental Concepts of Psychoanalysis*, trans. Alan Sheridan (New York & London: W.W. Norton & Co., 1998), 95.
28 Lacan, *Four Fundamental Concepts*, 95–96.
29 T.J. Clark, *The Sight of Death: An Experiment in Art Writing* (New Haven & London: Yale University Press, 2006), 2.
30 Clark, *The Sight of Death*, 8.
31 Lacan, *Four Fundamental Concepts*, 84.
32 Lacan, *Four Fundamental Concepts*, 88.
33 Lacan, *Four Fundamental Concepts*, 88.
34 Lacan, *Four Fundamental Concepts*, 88.
35 Susan Foister, 'Holbein's Extraordinary *Ambassadors*,' The National Gallery, London. https://www.youtube.com/watch?v=paA8hqqQ-_w [Accessed 24 September 2020].
36 Lacan, *Four Fundamental Concepts*, 89.
37 Harry Berger, Jr, *Caterpillage: Reflections on Seventeenth-Century Still Life Paintings* (New York: Fordham University Press, 2011), 68.
38 Hanneke Grootenboer, *The Rhetoric of Perspective: Realism and Illusionism in Seventeenth-Century Dutch Still-Life Painting* (Chicago & London: The University of Chicago Press, 2005), 5.
39 Elizabeth Alice Honig, 'Desire and Domestic Economy,' *The Art Bulletin*, Vol. 83, No. 2 (June 2011), 294.
40 Hans-Joachim Raupp, *Stillleben und Tierstücke* (Münster: LIT Verlag, 2004), 4. See also Svetlana Alpers, *The Art of Describing: Dutch Art in the Seventeenth Century* (London & Chicago: University of Chicago Press, 1983), xxi.
41 Alpers, *The Art of Describing*, 71.
42 Miya Tokumitso, 'The Currencies of Naturalism in Dutch "Pronk" Still Life Painting: Luxury, Craft, Envisioned Affluence,' *RACAR: revue d'art canadienne/Canadian Art Review*, Vol. 41, No. 2, The Nature of Naturalism: A Trans-Historcial Examination/La Nature du naturalisme: un questionnement transhistorique, 2016, 35.
43 Roland Barthes, *Critical Essays*, trans. Richard Howard (Evanston: Northwestern University Press, 1972), 5.

44 Barthes, *Critical Essays*, 5. See also Alpers, *The Art of Describing*, 75.
45 Julie Berger Hochstrasser, *Still Life and Trade in the Dutch Golden Age* (New Haven & London: Yale University Press, 2007), 9 & 257.
46 Roland Barthes, *Mythologies*, trans. Annette Leavers (New York: Farrar, Strauss & Giroux, 1972), 128, 116 & 122.
47 Barthes, *Critical Essays*, 6.
48 Hal Foster, 'The Art of Fetishism,' in *The Art of Fetishism*, ed. Emily Apter & William Pietz (Ithaca & London: Cornell University Press, 1993), 253.
49 Foster, 'The Art of Fetishism,' 254.
50 Foster, 'The Art of Fetishism,' 262.
51 See Freud, Sigmund. 'On Fetishism,' in *The Complete Psychological Works of Sigmund Freud*, Vol. XXI, trans. & ed., James Strachey (London: Hogarth & the Institute of Psychoanalysis, 1950), 198–204.
52 Foster, 'The Art of Fetishism,' 264.
53 Bryson, *Looking at the Overlooked*, 104 & 105.
54 Grootenboer, *The Rhetoric of Perspective*, 21.
55 Bryson, *Looking at the Overlooked*, 127.
56 Bryson, *Looking at the Overlooked*, 128.
57 Berger, Jr, *Caterpillage*, 34.
58 Neri, *The Insect and the Image*, xiii.
59 Bryson, *Looking at the Overlooked*, 122.
60 Norman Land, 'Giotto's Fly, Cimabue's Gesture, and "A Madonna and Child" by Carlo Crivelli,' *Source: Notes in the History of Art*, Vol. 15, No. 4 (Summer 1996), 13.
61 Connor, *Fly*, 96.
62 Chatwin, *Utz*, 35.
63 Deleuze & Guattari, *A Thousand Plateaus*, 321.
64 Connor, *Fly*, 15.
65 Chatwin, *Utz*, 35.
66 Connor, *Fly*, 13.
67 Chatwin, *Utz*, 35.
68 Jacques Derrida, 'Fourmis,' in *Hélène Cixous Rootprints: Memories of Life & Writing*, ed. Hélène Cixous & Mireille Calle-Gruber, trans. Eric Prenowitz (London & New York: Routledge, 1997), 121.
69 Michel Serres, *The Parasite*, trans. Lawrence R. Schehr (London & Minneapolis: University of Minnesota Press, 2007), 3.
70 Land, 'Giotto's Fly,' 13.
71 Land, 'Giotto's Fly,' 13.
72 Blanchot, *The Space of Literature*, 32.
73 Roger Caillois, 'Mimicry and Legendary Psychasthenia,' trans. John Shepley, *October*, Vol. 41, Winter 1984, 30.
74 Lacan, *Four Fundamental Concepts*, 99.
75 Caillois, 'Mimicry and Legendary Psychasthenia,' 20.
76 Caillois, 'Mimicry and Legendary Psychasthenia,' 28.
77 Caillois, 'Mimicry and Legendary Psychasthenia,' 30.
78 Lacan, *Four Fundamental Concepts*, 104.
79 Lacan, *Four Fundamental Concepts*, 74.
80 Bryson, *Looking at the Overlooked*, 13.
81 Grootenboer, *Rhetoric of Perspective*, 25.
82 Grootenboer, *Rhetoric of Perspective*, 77.
83 Bryson, *Looking at the Overlooked*, 61.
84 George Kubler cited in Bryson, *Looking at the Overlooked*, 138.
85 Bryson, *Looking at the Overlooked*, 139.

86 See Ingvar Bergström, *Dutch Still-Life Painting in the Seventeenth Century*, trans. Christina Hedström & Gerald Taylor (London: Faber & Faber Ltd., 1956).
87 Victor I. Stoichita, *The Self-Aware Image: An Insight into Early Modern Meta-Painting*, trans. Anne-Marie Glasheen (Cambridge, New York & Melbourne: Cambridge University Press, 1997), 20.
88 Jacques Derrida, *The Truth in Painting*, trans. Geoff Bennington & Ian McLeod (Chicago & London: The University of Chicago Press, 1987), 54.
89 Derrida, *The Truth in Painting*, 56.
90 Stoichita, *The Self-Aware Image*, 23.
91 Derrida, *The Truth in Painting*, 54.
92 Derrida, *The Truth in Painting*, 73.
93 Stoichita, *The Self-Aware Image*, 7.
94 Serres, *Parasite*, 38.
95 Miller cited in Stoichita, *The Self-Aware Image*, 24.
96 Serres, *Parasite*, 38.
97 Foucault, 'The Thought from the Outside,' 16 & 21.

8 Coda

Wolfman, Vienna, Larva

In *Seminar X: Anxiety* (1962–1963), Jacques Lacan refers to Freud's case study of the Wolfman and his famous dream—white wolves sitting still in a tree outside the window; it is winter, the landscape is covered in snow, the tree denuded—to discuss the 'arborification' of the subject, arrested in a state of fascination. Lacan writes:

> In the revelation of what appears to the Wolf Man through the gap and the frame—pre-figuring what I turned into a function—of the open window, and which can be identified in its form with the function of fantasy in its most anxiety-provoking mode, where lies the crux? Clearly it doesn't lie in the fact of knowing where the phallus is. It is, as it were, everywhere—identical to what I call the catatonia of the image of the tree and the perched wolves that [...] hold the subject in their gaze. There's no need to go looking for it in the five furry tails of the five animals. It is there in the very reflection of the image, which it supports with a catatonia that is nothing but the subject's own, the child turned to stone by what he sees, paralysed by this fascination to the point that one may conceive of what gazes back at him in the scene, and which is invisible on account of being everywhere, as nothing but the transposition of the arrested state of his own body, here transformed into a tree, *L'arbre couvert de loups* [...].[1]

In this passage, Lacan identifies the correspondence between the fantasy of the subject and the phallus, that is, the fantasy of the phallic subject mesmerizing and arresting to stone. The phallus in this instance consists of tree plus perched wolves, 'the catatonia of the image' assembling these different elements, though Lacan misses or overlooks some of them: white wolf-machine, to borrow Deleuze and Guattari's notation, plugged into tree, standing in a white winter landscape. Freud's *Wolfman* (1918) is a study of sacrifice, as Lacan observes, and ultimately of being unable to sustain the fantasy of the phallic subject (by default white, heterosexual, male, etc.), instead time and again brought into association, through other dreams, with the insectile. What I'd like to do in these concluding

DOI: 10.4324/9781003322757-8

remarks is to reflect back on the *Un/form* of the subject implicated in the insectile, entangled, from the early enunciations of psychoanalysis, with the production of total form. The focal point of this coda is the image of the cocoon deployed in, as well as outwith, Freud's case study as a function of racialisation inscribed here in the apparatus of the white wolf.

I have, over the course of this book, spoken about 'events' linked to insects—swarming, hatching, buzzing, squirming, moulting, etc.—and their various (affective, psychic) dimensions in the formation and undoing of the subject without really bringing to attention processes of cocooning, even if I have discussed issues relating to such operations or structures. In Cherish Oteka's 2022 documentary film *The Black Cop*, Gamal 'G' Turawa, an ex-Metropolitan police officer, recalls the elaborate layers of defensive armouring that he had to maintain in his job, which involved a 'joke' of literally being painted white, with shoe whitener, by his peers acting out what G already knew: that he's the 'wrong colour' to fit in.[2] The systematic and internalised racist (and heterosexist) abuse, as well as the instrumentalization of G's black body in promotional campaigns for the police force—including an appeal against knife crime, in which G is asked to hold a knife to incentivise others to dump their weapons, at once acting as agent against, and read by a majoritarian white culture as perpetrator of, knife attacks—forced him to consider suicide: the persona he had been compelled to put together finally shattering into pieces. To his counsellor's crucial question, 'who defines you?,' G does not initially know what to answer, because it demands a confrontation with the elusive other, whose *Che vuoi*, in general, compels the subject to anxiously respond by crafting an identity it thinks the other wants. This other facing G is an ideal fiction of masculinity, less a praying mantis (Lacan's premise in *Seminar X*) than a white wolf whose racialising gaze reduces the subject G to 'a caterpillar in [a] cocoon' that he's 'trying to break out of'.

I am deliberately confusing two (or more) narratives, the case study of the Wolfman and G's account of his life in the police force, condemning him to occupy a 'static ontological space,'[3] that is, the cocoon of an alienating identity that consumes him. The reason I am aligning these two narratives is that shared trope of the cocoon in Freud's *Wolfman* and Oteka's *The Black Cop*, both of whom to a certain extent hold it at their centre. I use the cocoon to think about racialisation: even though it is a figure of speech, it articulates the sensation of being seized and restrained in a particular abjected, frequently 'epidermalised,' objecthood.[4] In his book on Frantz Fanon, David Marriott performs a similar manoeuvre in relation to the experience of vertigo, which Fanon describes in *The Wretched of the Earth* (1961): anti-colonial activists become 'possessed with a kind of vertigo' when they realise they have reached certain limits in their struggle against the colonial regime; the trauma of colonial oppression and the subject's resistances against it fling it 'into a zone of turbulence,' a 'virtually pathological dreamlike state where the sight of the other [the coloniser] induces vertigo'.[5] Marriott observes that the term, vertigo, articulates the reality of existence in the colony, 'namely the ways in which the subject feels itself to be commanded' while referring to the rupture at the level of the subject's self-awareness. The I is constrained and at the same time strains against the racialising order that defines it, so that vertigo is a 'sign of crisis' indicative of a subject, commanded by the colonising other, whose life is made impossible or, at the very least, rendered aporetic.[6]

The cocoon equally is sign of crisis, of a captured subject and at the same time a figure that weaves itself around the invocation of the white wolf as a stand-in for, and representative of, the Ideal-I, which has in this project up until now appeared as coleopteran fantasy. In these final stages, though, the perspective shifts in terms of the form of this Ideal-I, the fictional direction that the subject must take. For the most part using Freud's study as starting point and point of return—Wolfman emerged across this book—I want to think about the insectile as, yet again, *informe* and concurrently the form the non-conforming I is ensnared into by the catastrophic demand of the racialising other. Catastrophic because the subject can never adequately meet this demand, a situation that, as Lacan shows in *Seminar X*, integrates anxiety into 'the net' of subject formation but that becomes untenable if the other's gaze either rends the I's every subjective form apart or, conversely, absolutely and totally fixes the I, 'overdetermined from without,' as Fanon writes.[7] There are obviously differences here which I do not want to abrade: Freud's analysand is a white man, an aristocrat, but he's Russian, initiating his therapy at the cusp of World War I, and there are degrees of racialisation just as there are 'gradations of whiteness' and hierarchies of the so-called human in the discourse and logic of race.[8] The analysis, furthermore, takes place in Vienna, between the Russian subject and an Austrian/culturally German Jew born in Moravia, in what is now the Czech Republic; the politics of this particular historical time and space affecting not just their sessions but the entire mechanism of Freudian psychoanalysis.

What I propose to do, then, all the while trying not to flatten difference and by adopting elements of Wolfman's primal scene/dream, is to, one last time for now, work through the racialised rendition of the I as insectile, produced as such by the phallic apparatus of the white wolf. If the latter functions as the fantasy-subject that is invisible because unmarked and everywhere, the former occurs, in Freud's case study but also frequently beyond it—hence the reference to *The Black Cop* as just one such instance—as a systematized cocooning and arresting of the racialised subject as forever larval. Cocooning, then, is the dream-image by which racialisation is experienced in psychic reality, although the insectile, as Fanon describes, takes on many forms in the racialising regime. In *Black Skin, White Masks* (1952), Fanon engages with Sartre's writing on antisemitism, noting that 'the Jew can be unknown in his Jewishness' whereas the 'BLACK MAN' (emphasis original, here and below) is instantly noticed, his body secreting the 'pseudopodia' of race.[9] He writes:

> I move slowly in the world, accustomed now to seek no longer for upheaval. I progress by crawling. And already I am being dissected under white eyes, the only real eyes. I am *fixed*. Having adjusted to their microtomes, they objectively cut away slices of my reality. I am laid bare. I feel, I see in those white faces that it is not a new man who has come in, but a new kind of man, a new genus.[10]

The 'new genus' is something that crawls, is dissected, fixed or pinned: these might be metaphors, but they nonetheless give an account of the affective and material reality, under white eyes and amongst white bodies, that the racialised subject has to inhabit and endure. The insectile, as we have seen, is not necessarily movement and metamorphosis, but also static or 'eternal' according to racist and antisemitic logic, imagining an unchanging 'defective' creature on the one hand and an idealised *typus* on the other.

Because I want to return the focus to Freud's Vienna, I'm proposing that we read cocooning as indicative of a de/articulation of particular subjects as *Minusvarianten* [defective variants], to use Julius Tandler's word, deployed in the context of Viennese interwar eugenic programmes.[11] Tandler (1869–1936) was a contemporary of Freud's and Professor of Anatomy at the University of Vienna as well as, briefly, in 1918, undersecretary of public health and subsequently municipal councillor for welfare and social administration in Vienna. He's a more ambiguous figure than his term *Minusvariante* implies, however: he was a Jewish Socialist, one of the very few Jewish professors holding chair positions in Vienna which, for over a decade, was governed by Karl Lueger, acting as the city's mayor from 1897 to 1910 and the founder of the 'officially antisemitic' Christian Social Party; Lueger is, thus, considered to be the founder of political antisemitism.[12] Britta McEwen further reminds us that so-called positive eugenics—the distinction derives from Alfred Ploetz, who defined people according to their '"positive" or "negative" biological materials' and came up with the term *Rassenhygiene*—forms the basis of the modern Austrian (and European) welfare state, with its emphasis on education, childcare, housing and food, to a considerable extent shaped by Tandler.[13] These reforms responded to the collapse of the city—due to failed sanitation systems, malnutrition, the spread of Spanish influenza, tuberculosis and sexually transmitted diseases—in the wake of World War I; they were, as such, preventative and, like psychoanalysis, emancipatory measures.[14] That the welfare state 'bears the mark,' as McEwen observes, of eugenic reform is an often occluded or repressed fact, yet concepts like *Minusvarianten* or *Rassenhygiene* circulated in the political discourse of the First Austrian Republic and were promulgated even by those later targeted through extermination policies, already contained, after all, in the constitution of those very words.[15]

If *Minusvariante* functioned as part of an emancipatory agenda (however incompatible such an agenda seems to be with the term), it refers, in this coda, exclusively to the *informe*, that which is considered and produced as lesser and can be crushed, to refer once more to Georges Bataille, like a spider, earthworm or insect. I am invoking *Minusvariante* in conjunction with the trope of the cocoon to paint a portrait of interwar Vienna: the cocoon as signified of a system of racialisation in operation in (and beyond) Freud's time and his city. If I used Hardy at the start of this book to think though the constellation between form, formlessness, the insectile and fascination, the investigation in this coda, of Freud's study of the Wolfman, returns to a primal scene, as it were, of cocooning and the terrorizing presence of the white phallic subject. Freud's analysis concentrates on the nexus between insectility, phallic subjectivity, and the threat of castration; he does not spend much time on the cocoon, other than intimating, perhaps, that it is indicative of his patient's 'fixed determination [*Fixierung*] to remain ill'.[16] The connection between cocoon and illness exists latently, intertextually: Freud talks about fixation but also of isolation, comfort, being suspended, anticipating the propositions he develops in *Beyond the Pleasure Principle* (1920). The psyche, he writes, needs to keep excitation low and seeks to eliminate the unpleasurable; the pleasure principle, as such, is a 'principle of constancy,' an 'expression of inertia,' a shield or cocoon against stimuli.[17] To refuse to overcome illness after initial therapy sessions, which start to unravel defence mechanisms installed and sought recourse to for years, means to retain a level of comfort: 'first moments of release' occur but then the analysand, terrified, prevents further possibilities of change.[18] In what is to come, the cocoon, as already mentioned, is coded differently, in relation to the valuation of form and the logic of racialisation, an approach that can be traced, despite the largely unacknowledged presence of such traces, in Freud's study.

The Wolfman relates his early memories—memories from the 'first estate,' referring to the country house that his family occupied up until he was five years old—to Freud in the following 'bundle' of recollections:

> There was a particular picture book, which showed a picture of a wolf standing on its hind legs and stepping out. Whenever he set eyes on his picture he would start to scream furiously, fearing that the wolf would come and gobble him up. [...] At the same time he was also afraid of other animals, both large and small. On one occasion he was chasing after a lovely big butterfly with yellow-striped wings that had pointed tips, trying to catch it. (Probably a 'swallowtail'.) Suddenly he was seized by a dreadful fear of the creature and gave up his pursuit, screaming. He also experienced fear and disgust at the sight of beetles and caterpillars. And yet he was able to recall that at

> the same period he had tortured beetles and cut up caterpillars; horses also gave him an uncanny feeling. He would scream if a horse was beaten and once had to leave the circus for this reason. On other occasions he enjoyed beating horses himself.[19]

The butterfly is a cutting device, associated with the V of a woman's legs, opening and closing like the swallowtail's beating wings: the yellow-striped butterfly-woman assemblage, in its suggestive V allowing 'access to the genitals,'[20] functions as castration machine. In another dream he recounts, the fear of the butterfly—the creature itself concealing the memory of his nanny—confirms this threat of castration: Wolfman dreamt of a 'man tearing the wings of an asp'.[21] Asp, or *Espe* in German, is a mutilated wasp [*Wespe*], the w the casualty of the scissoring V, the castrating nanny-butterfly system. There are, of course, more elements to decode in the passage above, the gobbling up by the upright father-wolf, the violently conflicting response to horses, the torture of beetles, caterpillars and flies, which Wolfman mentions elsewhere and whose wings he would, in turn, pull off: sado-masochistic pursuits traced back, according to Freud, to Wolfman's homosexual longings, the desire to be gobbled up by daddy. Love is that which demands sacrifice, to yield yourself up to the other by relinquishing your most treasured 'possession': the phallus, standing in (or up) for an autonomous subject. When Freud first meets Wolfman, the latter is 'completely dependent,' incapable of leading a self-sustained existence and instead 'entrenched behind a position of submissive indifference'.[22] He is totally cut off from the world through a series of repetition compulsions, a compulsive piety and the maintenance of fantasy formations that are, as Freud puts it, 'regressive in more than one sense,' in that they force his patient to '[shrink] back from life,' into illness, and to 'fall back on the past,' into childhood, that is, a reality absolutely produced by an (inadequate) adult.[23] What Wolfman infers, hence, is that in order to be loved, he must be castrated, to renounce a part of his body essential for the construction of an autonomous subjectivity, equivalent, here and more generally, to normative (white) masculinity.

Given that Wolfman equates butterflies with girls and women, but beetles and caterpillars with boys, his name is a misnomer (he becomes known by that which haunts him). He is, in fact, beetle-boy or larval form, whose underdeveloped I unleashes against life forms (mirror images) which it perceives with fear and disgust, giving him an 'uncanny feeling': beetles, caterpillars, asps, whose mutilation/castration he performs.[24] For Freud, pathologizing homosexuality, the I's phobic responses are the result of Wolfman's homosexual repression, initiated by the 'narcissistic masculinity of [his] genitals,'[25] seeking to protect an I constantly under siege, unable to raise itself up. We are, however, able to supplement Freud's reading by emphasising other aspects of his case study, elements that he pushed to the margins, all the while noting, as I have argued across this book, the correlations between the insectile and the development of the I. This correlation can variously be interpreted as the subject's misrecognition as imago, its cocooning in 'comfortable' illness and defensive structures screening out the terror of an independent existence or as the result of processes of racialisation. The insectile, in this latter sense, as that which is discarded, devalued, reified and racialised, is the figure of the never-quite-subject always already and forever set up to fail.

In his article on whiteness and countertransference in *Wolfman*, Alfred J. López remarks that Freud's last footnote to the study is evident of his own repression regarding the agency of whiteness, or of its gaze, remembering Wolfman's most famous dream: white wolves pinning their immobile gaze on a terrified, fascinated subject. In this footnote, Freud buries reference to his patient's provenance, that is, Russia; patient and Jewish analyst thereby sharing a precarious position vis-à-vis German or 'Aryan' whiteness, bearing in mind the 'hierarchies of whiteness' in operation then, just prior to World War I, as much as now.[26] Some groups, as López reminds us, are recognised as whiter than others: Jews are racialised, as are Slavic populations, both of whom operate as ethnic others unable to be 'Germanized'.[27] Neither patient nor analyst, according to this logic, is able to measure up to the 'standard' of whiteness, a fact of which they are both very much aware: on the one hand, Wolfman, whose sick father is an insufficient identification figure, prompting him to turn towards his German teacher and to develop 'a passion for military life, for uniforms, weapons and horses'.[28] On the other, the Jewish analyst, who displaces the 'marks' of Jewishness—as an antisemitic, internalised category pertaining to race, gender and sexuality—onto others, namely women and gay men, as Sander L. Gilman argues.[29] Freud relegates this correspondence between analyst and analysand, in terms of their respective marginal positions in Vienna and Austro-German culture, to the periphery, from whence it nonetheless exudes a curious influence. At once buried and central, because its 'obscene' presence can also serve to read the entire case study in a different light, the footnote addresses that which Freud does not want, or cannot allow himself, to see: his own exclusion from normative whiteness which, along with hegemonic masculinity, is what Freud needs to imprint on his patient.[30]

Wolfman's identification processes and sexual development have, in Freud's heterosexist perspective, failed; the relationship between father and son is, hence, profoundly ambivalent. It is determined by a 'passivity'—being acted upon—and insufficiency that should, as López shows, have facilitated countertransference, considering Freud's similar position, in the eyes of the dominant culture, as always already falling short of, excluded from, and persecuted by that culture. As such, it is unsurprising that Wolfman is a larval subject, dreaming of insect-beings at his heels, which he cuts up in waking life: he 'saw himself riding a horse, pursued by a giant caterpillar' in one dream; in another, the devil, in an 'upright stance' and with an 'outstretched finger,' points to a giant snail, an 'exquisite symbol of female sexuality,' as Freud adds.[31] His dreams are informed by extraordinary memories of a father and son asleep somewhere on the estate and of that same estate's trees as 'completely white, completely covered in caterpillar cocoons'.[32] Freud interprets these memories as residues funnelling into Wolfman's wolf tree dream, in which whiteness is one of the determining (and, in Freud analysis as well as Lacan's response, neglected) aspects facing the dreaming boy, held in a motionless, utterly calm gaze. Never mind, for now, Freud's reading of the scene—he attributes the gaze to the boy and explains that the stillness of the scene is to be translated as violent motion—the experience here is clearly one of being riveted to the spot: it is scene of fascination. The agency, as mentioned earlier, is whiteness, an interpretation which Freud enables only obliquely through that last footnote, indicative of the return of the repressed, displaced into the subtext. This agency operates through the upright figure of the wolf, whose whiteness, though arriving from elsewhere, instils in the dreaming child the association with caterpillar cocoons and, more generally, with the insectile.

The context to the case study of the Wolfman is, as already mentioned, World War I: analyst and analysand initially meet in the years leading up to it, from 1910 to 1914, whereas the study itself, though completed in 1914, was not published until 1918. It therefore contains, as López suggests, lessons for a 'postcolonial psychoanalysis,' bearing in mind Freud and Wolfman's precariousness in relation to hegemonic whiteness, the taking place of the analysis in an antisemitic Vienna, and the rival imperial powers of Germany and Tsarist Russia, casting Wolfman, too, as *Fremdkörper* at odds with the already mythologised *Volkskörper* of the German *Reich*.[33] There are a number of other factors, however, beyond Freud's footnote—added in 1918 and referring to the aftermath of WWI, by which point Wolfman had lost everything, is totally deracinated (without home, fortune, or family relations, as Freud notes) but seems all the more stable because of this deracination—that furnish the study with the context of things to come. If the footnote is a hinge on which turns the subject's relation to whiteness, the figure of the wolf, as representative of white, phallic masculinity, is another, even if this particular discourse is yet to be fully mobilised by the Nazis and, particularly, Hitler, whose totem animal was the wolf, which he spun into a whole 'totemic system'.[34] The latter included his own alias (a.k.a. *Wolf*), the naming of a number of his dog's cubs and the designations of strategic military command centres in France and (what is now) Poland, known as *Wolfsschanze* (wolf's lair) and *Wolfsschlucht* (wolf's glen).[35] In *The Wolf Man: Graphic Freud* (2012), when the Nazis arrive in Vienna, indicated by St. Stephen's Cathedral, after the *Anschluss* on 13 March 1938, they are accompanied or announced by a wolf (or a German shepherd), the *Herrentier* associated with the Nazi *Herrenmensch* ('master race').[36]

At the risk of bestowing the case study with a historical context it does not have, I would nonetheless like to follow this association, between wolf and predatory, (proto)fascist masculinity, because it belongs to an interpretive scaffolding—an apparatus—that is already present prior to Hitler's rise to power. In *A Thousand Plateaus* (1980), Deleuze and Guattari, discussing the case study, criticise Freud for bringing a pack of wolves back to 'molar unities,' the 'dismal' singularity of the Father, but this 'reversion,' as they see it, is not Freud's alone.[37] A pack, multiplicity or mass, according to Klaus Theweleit, is not per se anti-molar (which Deleuze and Guattari acknowledge): there are masses that are 'strictly formed' rather than formless, metamorphosing into a 'single creature'.[38] Theweleit's extensive two-volume investigation *Male Fantasies* concentrates on the *Freikorps*, soldiers who refused to disband after WWI and instigated a reign of white terror, encoded as acts of (self and national) defence. A memoir by a member of the *Freikorps*, Peter von Heydebreck, is entitled *Wir Wehr-Wölfe* (1931), which Theweleit (and his translators) misspells as *Wir Werwölfe*, an orthographic mistake—a slip of the pen as if to highlight ideological correspondences—collapsing the notion of defence [*Wehr*] with the apparition of the mythical werewolf, which, as Chantal Bourgault du Coudray, repeating Theweleit's mistake, argues, is 'object of positive identification' for the *Freikorps* and, later, for the Nazis.[39]

This project of 'becoming-wolf' echoes throughout Nazi mythology, which consecrated a special place (quite literally) for the author of an even earlier work with the title *Der Wehrwolf*, written by Hermann Löns and published in 1910. Löns died in the trenches in 1914 and was buried in a military cemetery in Loivre—the circumstances of the burial remain unclear—but Hitler had Löns's putative remains exhumed in 1934 and reinterred in Germany, more specifically the Lüneburg Heath, itself a formidable repository of Nazi myths and the setting of many of Löns's writings.[40] (In Thomas Pynchon's *Gravity's Rainbow* (1973), Dominus Blicero launches his 00000 rocket from there, seeking to establish a 'Deathkingdom' on the moon;[41] after the war, from 1946 to 1950, Adolf Eichmann found refuge in the region under an assumed name).[42] *Der Wehrwolf* is set during the Thirty Years' War in Ödringen, a village in the Heath, and concerns one Wulf Harm, a local farmer to become stuff of legend who organises a militia to defend against marauding *Fremdvölker*:

> Our lieutenant [*Hauptmann*] is called Wulf, and he is a real wolf, because where he bites there are thirty-three holes. Therefore, I propose that we call ourselves the defence wolves [*Wehrwölfe*], and as a sign of where we have resisted infamy, we leave three axe strokes, one here, one there, and a third one in the cross. And no one shall know anything about it, other than ourselves [...], and whosoever opens his mouth about it shall hang between two mangy dogs [...] until you no longer know who stinks the most.[43]

The *Wehrwölfe* mark their territory through the *Wolfsangel*—the three axe strokes—a heraldic figure and, weirdly, given the context here, thought to have been a device used to catch wolves or maybe to stabilise walls;[44] it was also frequently deployed in forestry, to plot a boundary or as a symbol on silvicultural uniforms. Above all, though, it became a Nazi marker—and remains a marker of the far right—designating certain branches of the Hitlerjugend as well as some divisions of the SS.[45] With his novel and its *Blut und Boden Romantik*—which occurs as '*Gut* [property] *und Blut*' in the story[46]—Löns activates the *Wolfsangel* in a political sense, in that it begins to function in terms of a 'Germanic' resistance while interpellating and therefore preserving a particular masculinity 'proper' to this *Volk*: the *Wehrwolf*. The hailing of this subject-form is, no doubt, the crucial reason why the *Wolfsangel* was especially associated with the Hitlerjugend, for whom Löns's novel was required reading and who wore the symbol as an arm patch; the *Wolfsangel* was further, in 1982, utilised by the *Junge Front*, the youth wing of the neo-Nazi organisation *Volkssozialistische Bewegung Deutschlands*, continuing to put it to work as a token of, and stimulus for, fanatical and thanatoid ideological subject formation. In contemporary Erfurt, in training camps and under the flag of the neo-Nazi party *Der III. Weg*, featuring the head of a wolf surrounded by an oak wreath, the collapse of the democratic order is rehearsed: the wish fulfilment of the I, standing guard, ready to pounce, shaped into *Wehrwolf*.[47]

By 1914, the wolf might not yet have acquired all this emblematic weight, though Löns's novel was published by the time Wolfman underwent analysis. My intention is not to prove that either Freud or his analysand read or was aware of the novel, even if it was reasonably popular prior to its bestseller status, systematically generated, to promote that idealised subject form, by the Nazis after 1933.[48] Rather, I wanted to follow and expand upon López's suggestion that this 'dream' subject, the Wolfman terrorized by that which is upright, was (and is) produced and maintained as a cultural imperative with distinct links to fascism. As we have seen elsewhere, notably in the chapter on Ernst Jünger, the form of his dream subject frequently coincides with the insectile as fantasy of an armoured body. For Wolfman, though, rendered as beetle-child in relation to the big white wolf, the relationship to the insect body is one of racialised preterition which, as Gilman argues, latently informs the development of Freudian psychoanalysis as a whole.[49] It is not, then, as if Freud '[knew] nothing,' as Deleuze and Guattari claim, 'about the fascination exerted by wolves' but that he understood it all too well.[50] After all, he writes about the libidinal structures of molar multiplicities in *Group Psychology and the Analysis of the Ego* (1921) and of the subject's hypnosis and capitulation in light of such multiplicities, the loyalty to the pack being one of the reasons that fascist ideology seizes upon wolves as totemic creatures. That Freud can't interpret the wolf-machine—which Deleuze and Guattari also call a swarming machine—as anything other than molar multiplicity has to do with the world in which he exists, the context of WWI and the period leading up to and including WWII and the formatting of the subject according to an Ideal-I that excludes him, his embeddedness in (Austro-)German culture produced as eternally parasitic. The wolf-machine, in this sense, is Nazi machine or 'proper' phallic subject machine, and if 'wolfing' and the insectile are bound up, as Deleuze and Guattari suggest, they are not necessarily indicative of the molecular or the rhizomatic. There should be no automatic assumption, on our part, that becoming-wolf or becoming-insect is resistant to form, the premise, to recall, of this book, alert to the ways in which *Un/forms* of subjectivity occur in a milieu of fascination integrating, as if on a Möbius strip, the so-called human into relations with the insectile.

Notes

1 Jacques Lacan, *The Seminar of Jacques Lacan: Anxiety, Book X*, trans. A.R. Price, ed. Jacques-Alain Miller (Cambridge & Malden, MA: Polity Press, 2015), 260.
2 Cherish Oteka, 'The Black Cop,' *The Guardian*, 2022. Available via: https://www.theguardian.com/world/ng-interactive/2022/jan/19/the-black-cop-a-police-officers-story-of-racism-remorse-and-resistance-documentary [Accessed 27 January 2022].
3 Gillian Straker, 'Race for Cover: Castrated Whiteness, Perverse Consequences,' *Psychoanalytic Dialogues*, Vol. 14, No. 4 (2008), 418.
4 On 'epidermalization' see Frantz Fanon, *Black Skin, White Masks*, trans. Charles Lam Markmann (London: Polity Press, 2008), 84; Paul Gilroy, *Against Race: Imagining Political Culture Beyond the Color Line* (Cambridge, MA: Belknap, 2000), 46; Michelle Stephens, 'Skin, Stain and Lamella: Fanon, Lacan and Inter-Racializing the Gaze,' *Psychoanalysis, Culture & Society*, Vol. 23, No. 3 (2018), 311 & 321.
5 Frantz Fanon, *The Wretched of the Earth*, trans. Richard Philcox (New York: Grove Press, 2004), 79 & 89.
6 David Marriott, *Whither Fanon: Studies in the Blackness of Being* (Stanford: Stanford University Press, 2018), 42–43.
7 Frantz Fanon, *Black Skin, White Masks*, 87. Lacan writes about 'anxiety in the net of signifiers' in *Anxiety*, 3–15.
8 Alfred J. López, 'Who's Afraid of the Big White Wolf? Whiteness, Counter-Transference and Freud's *Wolfman*,' *Psychoanalysis, Culture & Society*, Vol. 9, No. 2 (2004), 187.
9 Fanon, *Black Skin, White Masks*, 87 & 92.
10 Fanon, *Black Skin, White Masks*, 87.
11 The first time I heard the term was without the contextual information to Tandler during a talk by Robert Parzer, 'Von der "Euthanasie" zum Holocaust: Kranke und behinderte Menschen als Opfer des ersten nationalsozialistischen Mordprogramms,' Digitale Veranstaltung des BORUSSEUM [Borussia Dortmund] zum Holocaust Gedenktag, Dortmund, 26 January 2022. Available via: https://www.youtube.com/watch?v=pViDLA7IjuE [Accessed 2 February 2022].
12 Cheryl A. Logan, *Hormones, Heredity and Race: Spectacular Failure in Interwar Vienna* (New Brunswick, NJ: Rutgers University Press, 2013), 146 & 148.

13 Britta I. McEwen, 'Welfare and Eugenics: Julius Tandler's *Rassenhygienische* Vision for Interwar Vienna,' *Austrian History Yearbook*, Vol. 41, April 2010, 170–173.
14 Wolfgang Maderthaner, 'Austro-Marxism: Mass Culture and Anticipatory Socialism,' *Austrian Studies*, Vol. 14, No. 1 (2006), Culture and Politics in Red Vienna, 30 & 22–23.
15 McEwen, 'Welfare and Eugenics,' 170–171.
16 Sigmund Freud, *The 'Wolfman' and Other Cases*, trans. Louise Adey Huish (London, etc.: Penguin Books, 2002), 208.
17 Sigmund Freud, 'Beyond the Pleasure Principle, Group Psychology and Other Works,' in *The Standard Edition of the Complete Psychological Works of Sigmund Freud*, Vol. XVIII, ed. James Strachey (London: Vintage, 2001), 9 & 36.
18 Freud, *The 'Wolfman,'* 208.
19 Freud, *The 'Wolfman,'* 213–214.
20 Freud, *The 'Wolfman,'* 290.
21 Freud, *The 'Wolfman,'* 292.
22 Freud, *The 'Wolfman,'* 205 & 208.
23 Freud, *The 'Wolfman,'* 252.
24 Freud, *The 'Wolfman,'* 214.
25 Freud, *The 'Wolfman,'* 310.
26 López, 'Who's Afraid of the Big White Wolf?,' 187.
27 López, 'Who's Afraid of the Big White Wolf,' 187.
28 Freud, *The 'Wolfman,'* 268.
29 Sander L. Gilman, *Freud, Race and Gender* (Princeton: Princeton University Press, 1993), 8–9.
30 López, 'Who's Afraid of the Big White Wolf,' 201–202.
31 Freud, *The 'Wolfman,'* 268.
32 Freud, *The 'Wolfman,'* 269.
33 López, 'Who's Afraid of the Big White Wolf,' 189. See also Sylvia Jaworska, 'Anti-Slavic Imagery in German Radical Nationalist Discourse at the Turn of the Twentieth Century: A prelude to Nazi Ideology?,' *Patterns of Prejudice*, Vol. 45, No. 1 (2011), 435–452.
34 'Totemic system' are Freud's words in *Totem and Taboo*, trans. James Strachey (London & New York: Routledge, 2004), 4.
35 See Chantal Bourgault du Coudray, *The Curse of the Werewolf: Fantasy, Horror and the Beast within* (London & New York: I.B. Taurus, 2006), 100 and also *Hitler und Der Wolf: Rassenwahn im Dritten Reich*, N24 Doku, 2017. Available via YouTube: https://www.youtube.com/watch?v=mJvjhNNuLL4 [Accessed 17 January 2022].
36 Richard Appignanesi & Sława Harasymowicz, *The Wolf Man: Graphic Freud* (London: SelfMadeHero, 2012), 146.
37 Gilles Deleuze & Félix Guattari, *A Thousand Plateaus: Capitalism and Schizophrenia*, trans. Brian Massumi (London & New York: Continuum, 2004), 30–31.
38 Klaus Theweleit, *Male Fantasies*, Vol. II, trans. Erica Carter & Chris Turner (Minneapolis: University of Minnesota Press, 2003), 4.
39 Bourgault du Coudray, *The Curse of the Werewolf*, 106.
40 See 'Wer liegt im Grab von Hermann Löns?,' *NDR*, 28 July 2015. Accessible via: https://www.ndr.de/geschichte/schauplaetze/Wer-liegt-im-Grab-von-Hermann-Loens,loens114.html [Accessed 17 January 2022]. Thanks to Andreas Kahrs for bringing this to my attention.
41 Thomas Pynchon, *Gravity's Rainbow* (London: Vintage, 2013), 857.

42 Karsten Krüger, 'Eichmann: Unerkannt im Heidedorf,' *NWZ Online*, 23 July 2010. Accessible via: https://www.nwzonline.de/hintergrund/eichmann-unerkannt-im-heidedorf_a_1,0,1784739085.html# [Accessed 18 January 2022].

43 Hermann Löns, *Der Wehrwolf: Eine Bauernchronik* (Hannover & Hameln: Adolf Sponholtz Verlag, 1996), 78. [My translation. All further translations mine.]

44 See 'Symbole und Errkennungszeichen: Die Wolfsangel,' *Belltower News*, 21 July 2018. Available via: https://www.belltower.news/die-wolfsangel-51314/ [Accessed 18 January 2022].

45 Jens-Christian Wagner, 'Ideologie in Symbolen: Was haben Wolfsangel, Germanenkult und Hermann Löns mit dem Nationalsozialismus zu tun?,' *Stiftung Niedersächische Gedenkstätten*. Available via: https://www.stiftung-ng.de/fileadmin/dateien/GBB/Aktuell/Aktuell_temporaer/2019/Loens-Rezeption.pdf [Accessed 18 January 2022].

46 Löns, *Der Wehrwolf*, 79. Wagner, in 'Ideologie in Symbolen,' further refers to the novel's '*Blut-und-Boden-Romantik*,' n.p.

47 See 'Rechtsextreme Sportvereine: Kinder für Propaganda mißbraucht,' *Sport Inside*, 30 August 2019. Accessible via: https://www1.wdr.de/fernsehen/sport-inside/rechtsextreme-sportvereine-der-dritte-weg-100.html and also 'Rassismus auf Deutschen Fußballplätzen: Wie Neonazis Vereine unterwandern,' *Sportschau*, 27 February 2020. Accessible via: https://www.youtube.com/watch?v=nXZpUdQiRcs [Accessed 27 May 2022].

48 See Roderick H. Watt, 'Wehrwolf or Werwolf? Literature, Legend, or Lexical Error into Nazi Propaganda,' *The Modern Language Review*, Vol. 87, No. 4 (Oct 1992), 882–883.

49 See López, 'Who's Afraid of the Big White Wolf,' 188 and also Sander Gilman, *Freud, Race and Gender* (Princeton: Princeton University Press, 1993).

50 Deleuze & Guattari, *A Thousand Plateaus*, 32.

Bibliography

Adorno, Theodor. 'Punctuation Marks,' trans. Shierry Weber Nicholson. n.d. Accessible via: http://ubu.com/papers/Adorno-Theodor-W-Punctuation-Marks.pdf [Accessed 9 May 2022].

Agamben, Giorgio. *Potentialities*, trans. Daniel Heller-Roazen (Stanford: Stanford University Press, 1999).

Allen, Graham. 'The Unempty Wasp's Nest: Kubrick's *The Shining*, Adaptation, Chance, Interpretation,' *Adaptation*, Vol. 8, No. 3 (March 2015), 361–371.

Allerseelen, 'Käfer Lied,' Track Number 7 on *Stirb und Werde* (Aorta, 1999). Lyrics accessed via Allerseelen Facebook page: https://www.facebook.com/allerseelen.allerseelen/posts/10153644563784739/ [Accessed 5 November 2019].

Allerseelen, 'Käfer Lied,' YouTube, 8 February 2014: https://www.youtube.com/watch?v=Cg2pLF49RTc [Accessed 5 November 2019].

Alpers, Svetlana. *The Art of Describing: Dutch Art in the Seventeenth Century* (London & Chicago: University of Chicago Press, 1983).

Anzieu, Didier. *The Skin Ego*, trans. Chris Turner (New Haven & London: Yale University Press, 1989).

Ashenden, Liana. '*Ada*'s Erotic Entomology,' *Nabokov Studies*, Vol. 6, No. 1 (2000), 129–148.

Ascher, Rodney. *Room 237* (IFC Films, 2012).

Appignanesi, Richard & Sława Harasymowicz, *The Wolf Man: Graphic Freud* (London: SelfMadeHero, 2012).

Attali, Jacques. *Noise: The Political Economy of Music*, trans. Brian Massumi (Minneapolis & London: University of Minnesota Press, 2006).

Badmington, Neil, 'Theorizing Posthumanism,' *Cultural Critique*, No. 53 Posthumanism, Winter 2003, 10–27.

Barker, Jennifer M. *The Tactile Eye: Touch and the Cinematic Experience* (Berkeley, Los Angeles & London: University of California Press, 2009).

Barthes, Roland. *Critical Essays*, trans. Richard Howard (Evanston: Northwestern University Press, 1972a).

Barthes, Roland. *Mythologies*, trans. Annette Lavers (New York: Farrar, Straus & Giroux, 1972b).

Barthes, Roland. 'The Grain of the Voice,' in *The Sound Studies Reader*, ed. Jonathan Sterne (London & New York: Routledge, 2012), 504–510.

Barthes, Rolans. *The Neutral*, trans. Rosalind E. Krauss & Dennis Hollier (New York: Columbia University Press, 2005).

Barthes, Roland. *The Pleasure of the Text*, trans. Richard Miller (New York: Hill & Wang, 1975).
Bataille, Georges. *Encyclopaedia Acephalica: Comprising the Critical Dictionary & Related Texts*, trans. Iain White (London: Atlas Press, 1995).
Beaumont, Matthew. 'Cutting up the Corpse: Agatha Christie, Max Ernst and Neo-Victorianism in the 1930s,' *LIT: Literature, Interpretation, Theory*, Vol. 20, No. 1–2 (2009), 12–26.
Bechdel, Alison. *Fun Home: A Family Tragicomic* (London: Jonathan Cape, 2006).
Belltower News, 'Symbole und Errkennungszeichen: Die Wolfsangel,' 21 July 2018. Available via: https://www.belltower.news/die-wolfsangel-51314/ [Accessed 18 January 2022]
Benthien, Claudia. *Skin: On the Cultural Border between Self and the World*, trans. Thomas Dunlap (New York: Columbia University Press, 2002).
Bergström, Ingvar. *Dutch Still-Life Painting in the Seventeenth Century*, trans. Christina Hedström & Gerald Taylor (London: Faber & Faber Ltd., 1956).
Berger, Harry, Jr. *Caterpillage: Reflections on Seventeenth-Century Still Life Paintings* (New York: Fordham University Press, 2011).
Berlant, Lauren. *The Queen of America Goes to Washington City: Essays on Sex and Citizenship* (Durham & London, Duke University Press, 1997).
Berlant, Lauren. 'Structures of Unfeeling: *Mysterious Skin*,' *International Journal of Politics, Culture & Society*, Vol. 28, No. 3 (2015), 191–213.
Bhabha, Homi K. *The Location of Culture* (London: Routledge, 2004).
Blanchot, Maurice. *The Space of Literature*, trans. Ann Smock (Lincoln & London: University of Nebraska Press, 1989).
Blanchot, Maurice. *The Infinite Conversation*, trans. Susan Hanson (Minneapolis & London: University of Minnesota Press, 2003).
Bois, Yve-Alain and Rosalind E. Krauss, *Formless: A User's Guide* (New York: Zone Books, 1997).
Braidotti, Rosi. 'Are Bugs to Nature as Chips to Culture? On Becoming-Insect with Deleuze|Guattari' in *Deleuzian Events: Writing|History*, ed. Hanjo Berressem and Leyla Haferkamp (Hamburg: LIT, 2009), 146–168.
Braidotti, Rosi. *Metamorphoses: Towards a Materialist Theory of Becoming* (London: Polity, 2002).
Bramesco, Charles. '*High Life* Review—Orgasmic Brilliance in Deepest Space with Robert Pattinson,' *The Guardian*, 10 Sept 2018 Accessed via: https://www.theguardian.com/film/2018/sep/10/high-life-review-robert-pattinson-claire-denis-sci-fi-drama-astronauts [Accessed on 28 September 2021].
Brown, Lori. 'Becoming-Animal in the Flesh: Expanding the Ethical Reach of Deleuze and Guattari's Tenth Plateau,' *PhaenEx: Journal of Existential and Phenomenological Theory and Culture*, Vol. 2, No. 2 (Fall/Winter 2007), 260–278.
Bryson, Norman. *Looking at the Overlooked: Four Essays on Still Life Painting* (London: Reaktion, 1990).
Burns, Anna. *Milkman* (London: Faber & Faber, 2018).
Burstein, Jessica. 'Waspish Segments: Lewis, Prosthesis, Fascism,' *Modernism/Modernity*, Vol. 4, No. 2 (1997), 139–164.
Byatt, A.S. *Angels and Insects* (London: Vintage, 1995).
Caillois, Roger. 'Mimicry and Legendary Psychasthenia,' trans. John Shepley, *October*, Vol. 31, 1984, 17–32.

Cameron, James. *The Terminator* (Orion Pictures, 1984).

Canetti, Elias. *Crowds and Power*, trans. Carol Stewart (New York: Continuum, 1973).

Cavarero, Adriana. 'Multiple Voices,' in *The Sound Studies Reader*, ed. Jonathan Sterne (London & New York: Routledge, 2012), 520–538.

Chatwin, Bruce. *Utz* (New York, London, Victoria, Ontario & Auckland: Viking, 1988).

Cheng, Joyce. 'Mask, Mimicry, Metamorphosis: Roger Caillois, Walter Benjamin and Surrealism in the 1930s,' *Modernism/Modernity*, Vol. 16, No. 1 (January 2009), 61–86.

Chion, Michel. *Audio-Vision: Sound on Screen*, trans. Claudia Gorbman (New York: Columbia University Press, 1994).

Chion, Michel. *Sound: An Acoulogical Treatise*, trans. James A. Steintrager (Durham & London: Duke University Press, 2016).

Chion, Michel. *The Voice in Cinema*, trans. Claudia Gorbman (New York: Columbia University Press, 1999).

Cixous, Hélène. *Reading with Clarice Lispector*, trans. Verena Andermatt Conley (Minneapolis: University of Minnesota Press, 1990).

Clark, T.J., *The Sight of Death: An Experiment in Art Writing* (New Haven & London: Yale University Press, 2006).

Claverie, Ezra. 'Weird Interpellation,' *New Formations*, Vol. 84, Autumn 2015, 216–264.

Collignon, Fabienne, 'The Politics of a Smile,' *New Formations*, Vol. 95, Jan 2019, 87–102.

Connor, Steven. 'As Entomate as Intimate Can Pinchably Be,' *Modernist Transactions*, University of Bristol, 30 June 2000. Accessible via: http://stevenconnor.com/insects.html [Accessed 9 May 2022].

Connor, Steven. 'Fascination, Skin & Screen,' *Critical Quarterly*, Vol. 40, No. 1 (April 1998), 9–24.

Connor, Steven. *Fly* (London: Reaktion, 2006).

Connor, Steven. 'Making Flies Mean Something,' *Beckett and Animals*, University of Reading, 26 September 2009. Accessible via: http://stevenconnor.com/flymean.html [Accessed 17 December 2020].

Connor, Steven. *The Book of Skin* (London: Reaktion, 2004).

Constable, Charlotte. 'Surfaces of Science Fiction: Enacting Gender and "Humanness" in *Ex Machina*,' *Film-Philosophy*, No. 22, Vol. 2 (2018), 281–301.

Cronenberg, David. *Naked Lunch* (Alliance Releasing, 1991).

Cronenberg, David. *The Fly* (20th Century Fox, 1986).

Crowley, Patrick & Paul Hegarty, ed., *Formless: Ways in and out of Form* (Oxford, etc.: Peter Lang, 2005).

DaCosta, Nia. *Candyman* (Universal Pictures, 2021).

Davis, Colin. *Critical Excess: Overreading in Derrida, Deleuze, Levinas, Žižek and Cavell* (Stanford: Stanford University Press, 2020).

Deleuze, Gilles & Félix Guattari, *A Thousand Plateaus: Capitalism and Schizophrenia*, transl. Brian Massumi (London & New York: Continuum, 2004).

Deleuze, Gilles. *Cinema I: The Movement Image*, trans. Hugh Tomlinson & Barbara Habberjam (Minneapolis: University of Minnesota Press, 1997).

Deleuze, Gilles. *Cinema II: The Time-Image*, trans. Hugh Tomlinson & Robert Galeta (Minneapolis: University of Minnesota Press, 1989).

Deleuze, Gilles. *Essays Critical and Clinical*, trans. Daniel W. Smith & Michael A. Greco (London & New York: Verso, 1998).

Deleuze, Gilles. *The Logic of Sense*, trans. Mark Lester & Charles Stivale (London: Athlone Press, 1990).

Derrida, Jacques. *Acts of Literature*, ed. Derek Attridge (New York & London: Routledge, 1992a).

Derrida, Jacques. 'Force of Law: The "Mystical Foundation of Authority",' in *Deconstruction and the Possibility of Justice*, ed. Drucilla Cornell, Michael Rosenfeld & David Gray Carlson (New York & London: Routledge, 1992b), 3–67.

Derrida, Jacques. 'Fourmis,' in *Hélène Cixous Rootprints: Memories of Life & Writing*, ed. Hélène Cixous & Mireille Calle-Gruber, trans. Eric Prenowitz (London & New York: Routledge, 1997), 119–127.

Derrida, Jacques. '*Geschlecht* II: Heidegger's Hand,' in *Deconstruction and Philosophy: The Texts of Jacques Derrida*, ed. J. Sallis (Chicago & London: The University of Chicago Press, 1988), 161–196.

Derrida, Jacques. *Points …: Interviews, 1974 – 1994*, ed. Elizabeth Weber, trans. Peggy Kamuf & others (Stanford: Stanford University Press, 1995).

Derrida, Jacques. *Positions*, trans. Alan Bass (Chicago & London: University of Chicago Press, 1981).

Derrida, Jacques. 'The Law of Genre,' trans. Avital Ronell, *Critical Inquiry*, Vol. 7, No. 1 (Autumn 1980), 55–81.

Derrida, Jacques. *The Truth in Painting*, trans. Geoff Bennington & Ian McLeod (Chicago & London: The University of Chicago Press, 1987).

Derrida, Jacques. 'The Voice that Keeps Silence,' in *The Sound Studies Reader*, ed. Jonathan Sterne (London & New York: Routledge, 2012), 495–503.

Derrida, Jacques. *Writing and Difference*, trans. Alan Bass (New York & London: Routledge, 2001).

Dewald, Madeleine & Oliver Lammert, *Vom Hirschkäfer zum Hakenkreuz* (Dock 43, 2002).

Dinnen, Zara & Sam McBean. 'The Face as Technology,' *New Formations*, No. 93, Summer 2017, 122–137.

Donegan, Moira. 'Gay as in Happy: On Maggie Nelson,' *n+1*, Issue 23, n.d.. Available via: https://nplusonemag.com/issue-23/reviews/gay-as-in-happy/ [Accessed 9 May 2022].

Dólar, Mladen. *A Voice and Nothing More* (Cambridge, MA & London: MIT Press, 2008).

du Coudray, Chantal Bourgault. *The Curse of the Werewolf: Fantasy, Horror and the Beast within* (London & New York: I.B. Taurus, 2006).

Elsaesser, Thomas & Malte Hagener, *Film Theory: An Introduction through the Senses* (New York & London: Routledge, 2010).

Ernst, Max. *Une Semaine de Bonté* (New York: Dover Publications, 1976).

Faber, Michel. *Under the Skin* (Edinburgh: Canongate, 2000).

Fanon, Frantz. *Black Skin, White Masks*, trans. Charles Lam Markmann (London: Polity Press, 2008).

Fanon, Frantz. *The Wretched of the Earth*, trans. Richard Philcox (New York: Grove Press, 2004).

Farocki, Harun. *Stillleben* (Berlin: Movimentio, 1997). Accessible via: https://www.youtube.com/watch?v=KQkDUnVvryI [Accessed 27 October 2021].

Farokhmanesh, Mohammad & Frank Geiger. *Kleine Germanen: Eine Kindheit in der Rechten Szene* (Little Dream Pictures GmbH, 2019).
Feuerstein, Anna. 'Seeing Animals on Egdon Heath: The Democratic Impulse of Thomas Hardy's *The Return of the Native*,' *19: Interdisciplinary Studies in the Long Nineteenth Century*, Vol. 26, 2018, 1–18.
Fink, Bruce. *The Lacanian Subject: Between Language and Jouissance* (Princeton: Princeton University Press, 1995).
Flusser, Vilém. *Vampyroteuthis Infernalis*, trans. Valentine A. Pakis (Minneapolis & London: University of Minnesota Press, 2012).
Foister, Susan. 'Holbein's Extraordinary *Ambassadors*,' The National Gallery, London, n.d.: https://www.youtube.com/watch?v=paA8hqqQ-_w [Accessed 24 September 2020].
Foster, Hal. 'The Art of Fetishism,' in *Fetishism as Cultural Discourse*, ed. Emily Apter & William Pietz (Ithaca & London: Cornell University Press, 1993), 251–265.
Foucault, Michel. 'Maurice Blanchot: The Thought from the Outside,' in *Foucault/ Blanchot*, trans. Brian Massumi (New York: Zone Books, 1987), 7–58.
Foucault, Michel. *The Order of Things: An Archeology of the Human Sciences* (London & New York: Routledge, 2002).
François, Stéphane. 'The Euro-Pagan Scene: Between Paganism and Radical Right', trans. Ariel Godwin, *Journey for the Study of Radicalism*, Vol. 1, No. 2 (2007), 35–54.
Franklin, Seb. 'The Context of Forms,' *World Picture*, Vol. 11, Summer 2016, 1–23.
Freud, Sigmund. 'Beyond the Pleasure Principle, Group Psychology and Other Works,' in *The Standard Edition of the Complete Psychological Works of Sigmund Freud*, Vol. XVIII, ed. James Strachey (London: Vintage, 2001a).
Freud, Sigmund. *Group Psychology and the Analysis of the Ego*, trans. James Strachey (London: The Hogarth Press, 1949).
Freud, Sigmund. *Leonardo Da Vinci: A Psychosexual Study of an Infantile Reminiscence*, trans. A.A. Brill (New York: Moffat, Yard and Company, 1916).
Freud, Sigmund. 'Medusa's Head,' in *Writings on Art and Literature*. trans. James Strachey (Stanford: Stanford University Press, 1997), 264–265.
Freud, Sigmund. 'On Beginning the Treatment,' in *The Standard Edition of the Complete Psychoanalytical Works of Sigmund Freud*, Vol. XII, trans. & ed. James Strachey & others (London: The Hogarth Press and the Institute of Psychoanalysis, 1958), 123–144.
Freud, Sigmund. 'On Fetishism,' in *The Complete Psychological Works of Sigmund Freud*, Vol. XXI, trans. & ed. James Strachey (London: Hogarth & the Institute of Psychoanalysis, 1950), 198–204.
Freud, Sigmund. *Totem and Taboo*, trans. James Strachey (London & New York: Routledge, 2004).
Freud, Sigmund. 'The Psychopathology of Everyday Life,' in *The Standard Edition of the Psychological Works of Sigmund Freud*, Vol. VI, trans. & ed. James Strachey & others (London: Vintage, 2001b).
Freud, Sigmund. *The 'Wolfman' and Other Cases*, trans. Louise Adey Huish (London, etc.: Penguin Books, 2002).
Friedkin, William. *Bug* (Lionsgate, 2006).
Fry, Paul H. 'Moving Van: The Neverland Veens of Nabokov's *Ada*,' *Contemporary Literature*, Vol. 26, No. 2 (Summer 1985), 123–139.

Gass, William H. 'The Order of Insects,' in *In the Heart of the Heart of the Country* (New York: Harper & Row, 1969), 318–322.

Gfrereis, Heike. 'Schreiben als Träumen: Die Einklebungen in Ernst Jüngers Tagebüchern und Manuskripten,' in *Freiheit: Jünger Studien* 7, ed. Georg Knapp (Tübingen: Attempto Verlag Tübingen, 2015), 37–72.

Gfrereis, Heike with Ellen Strittmatter & Sonja Lehmann, 'Die Ausstellung,' in *Ernst Jünger: Arbeiter am Abgrund*, ed. Karlheinz Bohrer, Heike Gfrereis, Helmuth Lethen & Stephan Schlak (Marbach am Neckar: Deutsche Schillergesellschaft, 201), 103–219.

Gilman, Sander L. *Freud, Race and Gender* (Princeton: Princeton University Press, 1993).

Gilroy, Paul. *Against Race: Imagining Political Culture Beyond the Color Line* (Cambridge, MA: Belknap, 2000).

Glazer, Jonathan. *Under the Skin* (StudioCanal, 2013).

Goodman, Steve. 'The Ontology of Vibrational Force,' *The Sound Studies Reader*, ed. Jonathan Sterne (London & New York: Routledge, 2012), 70–72.

Gorenstein, Dan. 'Entomologische Horizontverschmelzung: Ernst Jüngers Hermeneutik der Käfer,' in *Inhalt: Perspektiven einer Kategorie Non Grata im Philosophischen Diskurs*, ed. Daniel Alder, Markus Christen, Jeannine Hauser & Christoph Steier (Würzburg: Königshausen & Neumann, 2015), 169–188.

Grootenboer, Hanneke. 'The Pensive Image: On Thought in Jan Van Huysum's Still Life Paintings,' *Oxford Art Journal*, Vol. 34, No. 1 (2011), 13–30.

Grootenboer, Hanneke. *The Rhetoric of Perspective: Realism and Illusionism in Seventeenth-Century Dutch Still-Life Painting* (Chicago & London: The University of Chicago Press, 2005).

Hainge, Greg. *Philippe Grandrieux: Sonic Cinema* (New York, London, Oxford, New Delhi & Sydney: Bloomsbury, 2017).

Hamacher, Werner. 'Working through Working,' trans. Matthew T. Hartman, *Modernism/Modernity*, Vol. 3, No. 1 (January 1996), 23–56.

Haraway, Donna. *Simians, Cyborgs and Women: The Reinvention of Nature* (New York: Routledge, 1991).

Hardy, Thomas. *The Return of the Native* (Cambridge, UK: Cambridge University Press, 2021).

Harman, Graham. *Weird Realism: Lovecraft and Philosophy* (Winchester & Washington: Zero Books, 2012).

Harris, Oliver. *Williams Burroughs and the Secret of Fascination* (Carbondale & Edwardsville: Southern Illinois University Press, 2003).

H.D., *Tribute to Freud* (Manchester: Carcanet, 1985).

Heise, Ursula K. *Imagining Extinction: The Cultural Meanings of Endangered Species* (Chicago: University of Chicago Press, 2016).

Herzog, Dagmar. 'Desire's Politics: Félix Guattari and the Renewal of the Psychoanalytic Left,' *Psychoanalysis and History*, Vol. 18, No. 1 (2016), 7–37.

Hilderbrand, Lucas. 'On the Matter of Blackness in *Under the Skin*,' *Jump Cut: A Review of Contemporary Media* n.d.. Accessibe via: https://www.ejumpcut.org/archive/jc57.2016/-HilderbrandUnderSkin/index.html [Accessed 10 October 2019].

Hitler und Der Wolf: Rassenwahn im Dritten Reich, N24 Doku, 2017. Available via YouTube: https://www.youtube.com/watch?v=mJvjhNNuLL4 [Accessed 17 January 2022].

Hochstrasser, Julie Berger. *Still Life and Trade in the Dutch Golden Age* (New Haven & London: Yale University Press, 2007).

Hollingsworth, Cristopher. *Poetics of the Hive: Insect Metaphor in Literature* (Iowa City: University of Iowa Press, 2001).

Honig, Elizabeth Alice. 'Desire and Domestic Economy,' *The Art Bulletin*, Vol. 83, No. 2, (June 2011), 294–315.

Hustak, Carla & Natasha Myers, 'Involutionary Momentum: Affective Ecologies and the Sciences of Plant/Insect Encounters,' *differences*, Vol. 23, No. 3 (2012), 74–118.

Huyssen, Andreas. 'Fortifying the Heart—Totally: Ernst Jünger's Armored Texts,' *New German Critique*, No. 59 Special Issue on Ernst Jünger, 1993: 3–23.

Jabr, Ferria. 'How Does a Caterpillar turn into a Butterly,' *Scientific American*, 10 August 2012. Accessible via: https://www.scientificamerican.com/article/caterpillar-butterfly-metamorphosis-explainer/ [Accessed 10 July 2020].

Jacobs, Nicole A. *Bees in Early Modern Transatlantic Literature: Sovereign Colony* (New York: Routledge, 2021).

Jaworska, Sylvia. 'Anti-Slavic Imagery in German Radical Nationalist Discourse at the Turn of the Twentieth Century: A prelude to Nazi Ideology?,' *Patterns of Prejudice*, Vol. 45, No. 1 (2011), 435–452.

Jünger, Ernst. *Coleopteren I*, 09.03.1961–12.12.1961, *Deutsches Literaturarchiv Marbach*, Marbach, Germany, n.d..

Jünger, Ernst. 'Der Arbeiter,' in *Sämtliche Werke* 8, Der Arbeiter (Stuttgart: Klett-Cotta, 1981a).

Jünger, Ernst. *The Worker*, trans. Bogdan Costea & Laurence Paul Hemming (Evanston: Northwestern University Press, 2017).

Jünger, Ernst, 'Typus, Name, Gestalt,' in *Sämtliche Werke* 13 (Stuttgart: Klett-Cotta, 1981b), 85–173.

Jünger, Ernst, 'Prometheus—Von Der Physik belehrt? Zu Wolfgang Wiesers Besprechung von *Typus, Name, Gestalt*,' in *Sämtliche Werke* 13 (Stuttgart: Klett-Cotta, 1981c), 174.

Jünger, Ernst. 'Total Mobilization,' in *The Heidegger Controversy: A Critical Reader*, ed. Richard Wolin (Cambridge, MA & London: MIT Press, 1993), 119–139.

Jünger, Ernst. 'Die Totale Mobilmachung,' in *Sämtliche Werke* 7, Betrachtungen Zur Zeit (Stuttgart: Klett-Cotta, 1980), 119–142.

Jünger, Ernst. 'Subtile Jagden,' in *Sämtliche Werke* 10 (Stuttgart: Klett-Cotta, 1978), 11–277.

Kahn, Douglas. *Noise, Water, Meat: A History of Sound in the Arts* (Cambridge, MA & London: MIT Press, 2001).

King, Stephen. *The Shining* (London: Hodder & Stoughton, 1990).

Kirkland, Isabella. *Taxa* (1999–2004), n.d. Accessible via: https://www.isabellakirkland.com/ [Accessed 9 May 2022].

Kittler, Friedrich. *Gramophone, Film, Typewriter*, trans. Geoffrey Winthrop-Young & Michael Wutz (Stanford: Stanford University Press, 1999).

Krauss, Rosalind E. *The Optical Unconscious* (Cambridge, MA: MIT Press, 1994).

Krüger, Karsten. 'Eichmann: Unerkannt im Heidedorf,' *NWZ Online*, 23 July 2010.Accessiblevia:https://www.nwzonline.de/hintergrund/eichmann-unerkannt-im-heidedorf_a_1,0,1784739085.html# [Accessed 18 January 2022].

Kubrick, Stanley. *2001: A Space Odyssey* (Metro-Goldwyn-Meyer, 1968).

Kubrick, Stanley. *The Shining* (Warner Bros., 1980).

Kuzma, Joseph D. *Maurice Blanchot and Psychoanalysis* (Leiden & Boston: Brill Rodopi, 2019).

Lacan, Jacques. *Four Fundamental Concepts of Psychoanalysis*, trans. A. Sheridan (New York & London: W.W. Norton & Co., 1998).

Lacan, Jacques. *The Seminar of Jacques Lacan: The Ego in Freud's Theory and in the Technique of Psychoanalysis, Book II*, ed. Jacques-Alain Miller, trans. Sylvana Tomaselli (Cambridge, New York, New Rochelle, Melbourne & Sydney: Cambridge University Press, 1988).

Lacan, Jacques. *The Seminar of Jacques Lacan: Transference, Book VIII*, ed. Jacques-Alain Miller, trans. Bruce Fink (Cambridge & Malden: Polity, 2015a).

Lacan, Jacques. *The Seminar of Jacques Lacan, Book X: Anxiety*, ed. Jacques-Alain Miller, trans. A.R. Price (Cambridge & Malden: Polity Press, 2015b).

Lacan, Jacques. 'The Mirror Stage as Formative of the Function of the I as Revealed in Psychoanalytic Experience,' in *Écrits: A Selection*, trans. Alan Sheridan (New York & London: Routledge, 2001), 1–6.

Lacaux, André. 'Blanchot et Lacan,' *Essaim*, Vol. 1, No. 14 (2005), 41–68.

Land, Norman. 'Giotto's Fly, Cimabue's Gesture, and "A Madonna and Child" by Carlo Crivelli,' *Source: Notes in the History of Art*, Vol. 15, No. 4 (Summer 1996), 11–15.

Lang, Fritz. *The Testament of Dr Mabuse* (Nero Film, 1933).

Lattanzio, Ryan. 'Mica Levi on Why Composing *Under the Skin* Was "Really Mental",' *Indie Wire*, 10 November 2014. https://www.indiewire.com/2014/11/mica-levi-on-why-composing-under-the-skin-was-really-mental-190232/ [Accessed 10 October 2019].

Leslie, Esther, ed. *On Photography* (London: Reaktion, 2016).

Lethem, Jonathan. *Motherless Brooklyn* (New York: Vintage, 2000).

Lethen, Helmut. *Cool Conduct: The Culture of Distance in Weimar Germany*, trans. Don Reneau (Berkeley, LA & London: University of California Press, 2002).

Levinas, Emmanuel. *Ethics and Infinity*, trans. Richard A. Cohen (Pittsburgh: Duquesne University Press, 1985).

Levine, Caroline. *Forms: Whole, Rhythm, Hierarchy, Network* (Princeton & Oxford: Princeton University Press, 2015).

Lispector, Clarice. *The Passion According to G.H.*, trans. Idra Novey (London, etc.: Penguin Books, 2014).

LoBrutto, Vincent. *Stanley Kubrick: A Biography*, (New York, London, Victoria, Toronto & Auckland: Donald I. Fine Books, 1997).

Logan, Cheryl A. *Hormones, Heredity and Race: Spectacular Failure in Interwar Vienna* (New Brunswick, NJ: Rutgers University Press, 2013).

Löns, Hermann. *Der Wehrwolf: Eine Bauernchronik* (Hannover & Hameln: Adolf Sponholtz Verlag, 1996).

López, Alfred J. 'Who's Afraid of the Big White Wolf? Whiteness, Counter-Transference and Freud's *Wolfman*,' *Psychoanalysis, Culture & Society*, Vol. 9, No. 2 (2004), 186–206.

Lovecraft, H.P. 'At the Mountains of Madness,' in *At the Mountains of Madness & Other Novels* (London: Gollancz, 1971a), 1–100.

Lovecraft, H.P. 'The Case of Charles Dexter Ward,' in *At the Mountains of Madness & Other Novels* (London: Gollancz, 1971b), 101–221.

Lovecraft, H.P. 'The Dreams in the Witch-House,' in *At the Mountains of Madness & Other Novels* (London: Gollancz, 1971c), 248–283.

Lovecraft, H.P. 'The Whisperer in the Darkness,' in *The Whisperer in the Darkness and Other Stories* (London: Wordsworth, 2007), 217–276.

Lovecraft, H.P. 'The Shadow over Innsmouth,' in *The Lurking Fear and Other Stories* (London: Panther Books, 1964a), 103–155.

Lovecraft, H.P. 'The Shadow out of Time,' in *The Lurking Fear and Other Stories* (London: Panther Books, 1964b), 155–208.

Lovecraft, H.P. 'The Shunned House,' in *At the Mountains of Madness & Other Novels* (London: Gollancz, 1971d), 222–247.

Luckhurst, Roger. 'Corridor Dread: A History of Institutional Fear,' *The Wellcome Collection*, n.d. Available from: https://soundcloud.com/wellcomecollection/corridor-dread-a-history-of-institutional-fear [Accessed 10 October 2019].

Luckhurst, Roger. *The Shining* (Basingstoke & New York: Palgrave Macmillan, 2013).

Luckhurst, Roger. 'The Weird: A Disorientation,' *Textual Practice*, Vol. 31, No. 6 (2017), 1041–1061.

Maderthaner, Wolfgang. 'Austro-Marxism: Mass Culture and Anticipatory Socialism,' *Austrian Studies*, Vol. 14, No. 1 Culture and Politics in Red Vienna, (2006), 21–36.

Maeterlink, Maurice. *The Life of the Ant*, trans. Bernard Miall (London: George Allen and Unwin, Ltd., 1941).

Marks, Laura. *The Skin of the Film: Intercultural Cinema, Embodiment and the Senses* (Durham, NC: Duke University Press, 2000).

Marriott, David. *Whither Fanon: Studies in the Blackness of Being* (Stanford: Stanford University Press, 2018).

Maude, Ulrika. '"A Stirring Beyond Coming and Going": Beckett and Tourette's,' *Journal of Beckett Studies*, Vol. 17, No. 1–2 (September 2008), 153–168.

Marx, Karl. *Capital*, Vol. I, trans. Ben Fowkes (London, etc.: Penguin Books, 1990).

McEwen, Britta I. 'Welfare and Eugenics: Julius Tandler's *Rassenhygienische* Vision for Interwar Vienna,' *Austrian History Yearbook*, Vol. 41, April 2010, 170–190.

Meynen, Gloria. '/ [Über den Schrägstrich],' in *Satzzeichen: Szenen der Schrift*, ed. Helga Lutz, Nils Plath & Dietmar Schmidt (Berlin: Kadmos, 2017), 143–147.

Miéville, China & Jeffrey Andrew Weinstock. 'Afterword: Interview with China Miéville,' in *The Age of Lovecraft*, ed. Carl H. Sederholm & Jeffrey Andrew Weinstock (Minneapolis: University of Minnesota Press, 2016), 231–243.

Miéville, China. 'M.R. James and the Quantum Vampire: Weird; Hauntological: Versus and/or and and/or or?,' *Weird Fiction Review*, 29 November 2011. Accessible via: http://weirdfictionreview.com/2011/11/m-r-james-and-the-quantum-vampire-by-china-mieville/ [Accessed 9 May 2022].

Mitchell, W.J.T. *What Do Pictures Want: The Lives and Loves of Images* (Chicago & London: The University of Chicago Press, 2005).

Murray, Rachel. *The Modernist Exoskeleton: Insects, War, Literary Form* (Edinburgh: Edinburgh University Press, 2020).

Nabokov, Vladimir. *Ada or Ardor* (London, etc.: Penguin Books, 2015).

Nabokov, Vladimir. *Speak, Memory: An Autobiography Revisited* (London: Weidenfeld & Nicholson, 1967).

Nancy, Jean-Luc. *Listening*, trans. Charlotte Mandell (New York: Fordham University Press, 2007). Print.
Norddeutscher Rundfunk [NDR], 'Wer liegt im Grab von Hermann Löns?,' 28 July 2015. Accessible via: https://www.ndr.de/geschichte/schauplaetze/Wer-liegt-im-Grab-von-Hermann-Loens,loens114.html [Accessed 17 January 2022].
Nedoh, Boštjan & Andreja Zevnik, ed. *Lacan and Deleuze: A Disjunctive Synthesis* (Edinburgh: Edinburgh University Press, 2017).
Neri, Janice. *The Insect and the Image: Visualising Nature in Early Modern Europe, 1500–1700* (Minneapolis: University of Minnesota Press, 2011).
Ngai, Sianne. *Ugly Feelings* (Cambridge, MA & London: Harvard University Press, 2007).
Oates, Joyce Carol. *The Sacrifice* (London: Fourth Estate, 2015).
Ockenden, Amy. 'Seeing the Unseen: Investigating the Polarisation Properties of the Scarab Beetle Exocuticle and the Adaptive Role of reflecting Polarised Light,' Unpublished Practical Project, University of Auckland, 2 January 2017. Accessible via: https://www.researchgate.net/publication/315410121_Investigating_the_polarisation_properties_of_the_scarab_beetle_exocuticle_and_the_adaptive_role_of_reflecting_polarised_light [Accessed 19 November 2021].
O'Connor, Frank. *The Lonely Voice: A Study of the Short Story* (New York: Melville House Publishing, 2004).
Ogilvie, Brian W. 'Order of Insects,' in *The Life Sciences in Early Modern Philosophy*, ed. Ohad Nachtomy (Cary: Oxford University Press, 2013), 222–250.
Orange, Tommy. *There There* (London: Vintage, 2018).
Osterweil, Ara. '*Under the Skin*: The Perils of Becoming Female,' *Film Quarterly*, Vol. 67, No. 4 (Summer 2014), 44–51.
Oteka, Cherish. 'The Black Cop,' *The Guardian*, 2022. Available via: n.d.:https://www.theguardian.com/world/ng-interactive/2022/jan/19/the-black-cop-a-police-officers-story-of-racism-remorse-and-resistance-documentary [Accessed 27 January 2022].
Parikka, Jussi. *Insect Media: An Archaeology of Animals and Technology* (Minneapolis & London: University of Minnesota Press, 2010).
Parzer, Robert. 'Von der "Euthanasie" zum Holocaust: Kranke und behinderte Menschen als Opfer des ersten nationalsozialistischen Mordprogramms,' Digitale Veranstaltung des BORUSSEUM [Borussia Dortmund] zum Holocaust Gedenktag, Dortmund, 26 January 2022. Available via: https://www.youtube.com/watch?v=pViDLA7IjuE [Accessed 2 February 2022].
Peele, Jordan. *Get Out* (Universal Pictures, 2017).
Pettman, Dominic. *Creaturely Love: How Desire Makes Us More and Less Than Human* (Minneapolis & London: University of Minnesota Press, 2017).
Pietz, William. 'The Problem of the Fetish I,' *Anthropology and Aesthetics*, No. 9, Spring 1985, 5–17.
Poe, Edgar Allan. 'The Gold Bug,' in *The Complete Tales and Poems of Edgar Allan Poe* (London, etc.: Penguin Books, 1982), 42–70.
Pynchon, Thomas. *Gravity's Rainbow* (London: Vintage, 2013).
Quinlivan, Davinia. *The Place of Breath in Cinema* (Edinburgh: Edinburgh University Press, 2012).
'Rassismus auf Deutschen Fußballplätzen: Wie Neonazis Vereine unterwandern,' *Sportschau*, 27 February 2020. Accessible via: https://www.youtube.com/watch?v=nXZpUdQiRcs [Accessed 27 May 2022].

Raupp, Hans-Joachim. *Stillleben und Tierstücke* (Münster: LIT Verlag, 2004).
'Rechtsextreme Sportvereine: Kinder für Propaganda mißbraucht,' *Sport Inside*, 30 August 2019. Accessible via: https://www1.wdr.de/fernsehen/sport-inside/rechtsextreme-sportvereine-der-dritte-weg-100.html [Accessed 27 May 2022].
Reich, Wilhelm. *Character Analysis*, trans. Vincent R. Carfagno (New York: Farrar, Straus & Giroux, 1990).
Rendell, Jane. *The Architecture of Psychoanalysis: Spaces of Transition* (London & New York: I.B. Tauris, 2017).
Rose, Bernard. *Candyman* (TriStar Pictures, 1992).
Rose City Anti-Fascists, 'Austrian Far-Right Musical Project on West Coast Tour,' *Who Makes the Nazis?*, 14 December 2010. Accessible via: https://rosecityantifa.org/articles/allerseelen-on-tour/ [Accessed 9 May 2022].
Sartor, Genviève. 'A Joycean Interface: Re-Territorialising between Deleuze and Lacan,' *University of Toronto Quarterly*, Vol. 87, No. 2 (Spring 2018), 1–15.
Sartre, Jean-Paul. *Being and Nothingness*, trans. Hazel E. Barnes (New York, London, Toronto, Sydney & Singapore: Washington Square Press, 1992).
Scarry, Elaine. *The Body in Pain: The Making and Unmaking of the World* (Oxford: Oxford University Press, 1985a).
Schlak, Stefan. 'Ernst Jünger im Archive gelesen: Sechs Stereoskopien,' in *Ernst Jünger: Arbeiter am Abgrund*, ed. Karlheinz Bohrer, Heike Gfrereis, Helmuth Lethen & Stephan Schlak, (Marbach am Neckar: Deutsche Schillergesellschaft, 2010), 11–101.
Schleifer, Ronald. 'The Poetics of Tourette's Syndrome: Language, Neurobiology and Poetry,' *New Literary History*, Vol. 32, No. 3 (Summer 2001), 563–584.
Schulz, Ulrich K.T. *Der Hirschkäfer* (Berlin: Universum-Film AG (UFA), 1921).
Scarry, Elaine. *The Body in Pain: The Making and Unmaking of the World* (Oxford: Oxford University Press, 1985b).
Scott, Ridley. *Alien* (20th Century Fox, 1979).
Scott, Ridley. *Blade Runner* (Warner Bros., 1982).
Sears, John. *Stephen King's Gothic* (Cardiff: University of Wales Press, 2011).
Shekhovtsov, Anton. 'Apoliteic Music: Neo-Folk, Martial Industrial and "Metapolitical Fascism"', *Patterns of Prejudice*, Vol. 43, No. 5 (2009), 431–457.
Serres, Michel. *Genesis*, trans. Geneviève James & James Nielson (Ann Arbor: University of Michigan Press, 1999).
Serres, Michel. *The Parasite*, trans. Lawrence R. Schehr (London & Minneapolis: University of Minnesota Press, 2007).
Sleigh, Charlotte. *Six Legs the Better: A Cultural History of Myrmecology* (Baltimore: Johns Hopkins University Press, 2007).
Sobchack, Vivian. *Carnal Thoughts: Embodiment and Moving Image Culture* (Berkeley: University of California Press, 2004).
Sommer, Manfred. *Sammeln: Ein Philosophischer Versuch* (Frankfurt am Main: Suhrkamp, 1999).
Smith, Daniel. 'What Is the Body without Organs? Machine and Organism in Deleuze and Guattari,' *Contemporary Philosophy Review*, Vol. 51, 2018, 95–110.
Sperl, Stephan. *Die Semantisierung der Musik im filmischen Werk Stanley Kubricks* (Würzburg: Könighausen & Neumann, 2006).
Spillers, Hortense J. 'Mama's Baby, Papa's Maybe: An American Grammar Book,' *Diacritics* Vol. 17, No. 2 Culture and Counter-Memory: The 'American' Connection (Summer 1987). 64–81. Print.

Stephens, Michelle. 'Skin, Stain and Lamella: Fanon, Lacan and Inter-Racializing the Gaze,' *Psychoanalysis, Culture & Society*, Vol. 23, No. 3 (2018), 310–329.

Stoichita, Victor I. *The Self-Aware Image: An Insight into Early Modern Meta-Painting*, trans. Anne-Marie Glasheen (Cambridge, New York & Melbourne: Cambridge University Press, 1997).

Straker, Gillian. 'Race for Cover: Castrated Whiteness, Perverse Consequences,' *Psychoanalytic Dialogues*, Vol. 14, No. 4 (2008), 405–422.

Tartt, Donna. *The Goldfinch* (New York, Boston, London: Little, Brown and Co., 2013).

Tartt, Donna. *The Little Friend* (London: Bloomsbury, 2002).

Taylor, Sunaura. *Beasts of Burden: Animal and Disability Liberation* (New York & London: The New Press, 2016).

Thacker, Eugene. 'Pulse Demons,' *The Funambulist*, 13 January 2013. Accessible via:https://thefunambulist.net/philosophy/philosophy-pulse-demons-by-eugene-thacker [Accessed 29 April 2022].

Thacker, Eugene. 'Networks, Swarms, Multitudes,' Parts One and Two, *C-Theory*, n.d. Available via: http://ctheory.net/ctheory_wp/networks-swarms-multitudes-part-one/?template=print. http://ctheory.net/ctheory_wp/networks-swarms-multitudes-part-two/ [Accessed 10 October 2019].

The Guardian. 'Interview with Mica Levi,' 15 March 2014. https://www.theguardian.com/music/2014/mar/15/mica-levi-under-the-skin-soundtrack [Accessed 10 October 2019].

Theweleit, Klaus. *Male Fantasies*: *Psychoanalysing the White Terror*, Vol II, trans. Erica Carter & Chris Turner (Minneapolis: University of Minnesota Press, 1989).

Tokarczuk, Olga. *Flights*, trans. Jennifer Croft (New York: Riverhead Books, 2018).

Tokumitso, Miya. 'The Currencies of Naturalism in Dutch "Pronk" Still Life Painting: Luxury, Craft, Envisioned Affluence,' *RACAR: revue d'art canadienne/Canadian Art Review*, The Nature of Naturalism: A Trans-Historcial Examination/La Nature du naturalisme: un questionnement transhistorique, Vol. 41, No. 2 (2016), 30–43.

Truman, James W. 'The Evolution of Insect Metamorphosis,' *Current Biology*, Vol. 29, No. 3 (2019), R1252–R1268.

Tunbridge, Laura. 'Scarlett Johansson's Body and the Materialisation of the Voice,' *Twentieth-Century Music*, Vol. 13, No. 1 (2016), 139–152.

van Elferen, Isabella. 'HYPER-CACOPHONY: Lovecraft, Speculative Realism, and Sonic Materialism,' in *The Age of Lovecraft*, ed. Carl H. Sederholm & Jeffrey Andrew Weinstock (Minneapolis: U of Minnesota Press, 2016), 79–96.

Vehlken, Sebastian. *Zootechnologien: Eine Mediengeschichte der Schwarmforschung* (Zürich: diaphanes, 2012).

Verhoeven, Paul. *Starship Troopers* (Sony Pictures, 1997).

Vint, Sheryl. 'Skin Deep: Alienation in *Under the Skin*,' *Extrapolation*, Vol. 56, No. 1 (Spring 2015), 1–14.

Wagner, Jens-Christian. 'Ideologie in Symbolen: Was haben Wolfsangel, Germanenkult und Hermann Löns mit dem Nationalsozialismus zu tun?,' *Stiftung Niedersächische Gedenkstätten*, n.d. Available via: https://www.stiftung-ng.de/fileadmin/dateien/GBB/Aktuell/Aktuell_temporaer/2019/Loens-Rezeption.pdf [Accessed 18 January 2022].

Walton, Saige. *Cinema's Baroque Flesh: Film, Phenomenology and the Art of Entanglement* (Amsterdam: Amsterdam University Press, 2016).

Watson, Janell. *Guattari's Diagrammatic Thought: Writing Between Lacan and Deleuze* (London: Bloomsbury, 2009).

Watt, Roderick H. 'Wehrwolf or Werwolf? Literature, Legend, or Lexical Error into Nazi Propaganda,' *The Modern Language Review*, Vol. 87, No. 4 (October 1992), 879–895.

Weheliye, Alexander G. *Habeas Viscus: Racialising Assemblages, Biopolitics, and Black Feminist Theories of the Human* (Durham & London: Duke University Press, 2014).

Weinroth, Michelle. '"Morpho Eugenia" and the Fictions of Victorian Englishness: A.S Byatt's Postcolonial Critique,' *English Studies in Canada*, Vol. 31, No. 2–3 (June–September 2005), 187–222.

Wellmann, Janina. 'Die Metamorphose der Bilder,' *NMT: International Journal of History & Ethics of Natural Sciences, Technology and Medicine*, No. 16, No. 2 (2008), 183–211.

Whitehead, Colson. *Sag Harbor* (London: Vintage Books, 2011).

Wilcox, Lauren. 'Drones, Swarms, and Becoming-Insect: Feminist Utopias and Posthuman Politics,' *Feminist Review*, Vol. 116, No. 1 (2017), 25–45.

Williams, Claire. '*The Passion According to G.H.* by Clarice Lispector,' in *The Cambridge to the Latin American Novel*, ed. Efraín Kristal (Cambridge, UK: Cambridge University Press, 2005), 245–257.

Wills, David. *Dorsality: Thinking Back through Technology and Politics* (Minneapolis & London: University of Minnesota Press, 2008).

Wood, Michael. *The Magician's Doubts: Nabokov and the Risks of Fiction* (London: Chatto & Windus, 1994).

Woodard, Ben. 'Mad Speculation and Absolute Inhumanism: Lovecraft, Ligotti and the Weirding of Philosophy,' *Continent*, Vol.1, No. 1 (2011), 3–13.

Woolf, Virginia. *Three Guineas* (New York: Harcourt, Brace and Co., 1938).

Wynter, Sylvia. 'Unsettling the Coloniality of Being/Power/Truth/Freedom: Towards the Human, after Man, Its Overrepresentation—An Argument,' *CR: The Centennial Review*, Vol. 3, No. 3 (Fall 2003), 257–337.

Xu, Man, Ainsley E. Seago, Tara D. Sutherland, & Sarah Weisman, 'Dual Structural Colour Mechanisms in a Scarab Beetle,' *Journal of Morphology*, Vol. 271, No. 11 (2010), 1300–1305.

Žižek, Slavoj. *Looking Awry: An Introduction to Jacques Lacan through Popular Culture* (Cambridge, MA & London: October Books, 1991).

Index

For Product Safety Concerns and Information please contact our EU
representative GPSR@taylorandfrancis.com
Taylor & Francis Verlag GmbH, Kaufingerstraße 24, 80331 München, Germany

www.ingramcontent.com/pod-product-compliance
Ingram Content Group UK Ltd.
Pitfield, Milton Keynes, MK11 3LW, UK
UKHW020135200926
475702UK00005B/752
* 9 7 8 1 0 3 2 3 4 5 5 0 5 *